SECRETS OF ALCHEMY AND MYSTICAL SCIENCES

A Legendary Quest from Unknown Magic and Spirituality to the Modern Age of Chemistry

HARMINDER GILL

Printed in the United States of America
Library of Congress Control Number: 2024919331
ISBN: Softcover 979-8-89518-296-3
e-Book 979-8-89518-297-0
Hardback 979-8-89518-376-2
Published by: WP Lighthouse
Publication Date: 10/08/2024

To buy a copy of this book, please contact:
WP Lighthouse
Phone: +1-888-668-2459
support@wplighthouse.com
wplighthouse.com

TABLE OF CONTENTS

ACKNOWLEDGEMENTS

I wish to acknowledge WP Lighthouse for taking the time to publish the book. Thank you very much for the time, dedication, and the editing process for this book. It is with such great pleasure to have everyone on the team to make the book possible for many readers worldwide.

Chapter 1

FOUNDATIONAL CONCEPTS

Alchemy was an area of study that was rich with symbols that were more than signs. Alchemists use these symbols to include a unique blend of science, mysticism, and philosophy. They were also used as a universal language during ancient times and a way to guard their secretive practices of alchemy. There were beyond practical meanings of alchemical symbols that were held to be interesting symbolic representations. Major alchemical symbols include "The Three Primes". The concept of the Tria Prima or three primes was introduced by a 16th-century Swiss philosopher Paracelsus. He came up with the idea that these elements were needed to understand diseases, their cures, and defining human nature with each element that correlates to the identity of humans. These elements include mercury which symbolizes the mind and the state of transcending mortality. It was known as quicksilver. It was thought that quicksilver oscillated between liquid and solid that symbolizes the transition between life and death. The second element was considered to be salt which represents the body or physical matter. They felt it related to solidity, stability, and crystallization that symbolized the tangible physical reality of the world and the identity of human beings. The third element symbolized soul or emotions. It is associated with physical properties such as heat and dryness which mirrors the fiery aspects of the soul. Sulfur was thought of the driving force which is similar to how emotions can drive actions and reactions of human beings. These three primes were seen as the foundational substances and principles that make up the universe and everything including human beings.

The four classical elements was important to how alchemists viewed the universe and the nature of matter. Earth was considered to be an element that symbolizes solidity, stability, and physicality. In alchemy, Earth represents the physical body or a solid state of matter. It was associated with qualities of roundedness, heaviness, and permanence. It was considered to be the foundation or base which everything was built from. Air was the other element and symbolizes the intangible both the mental and spiritual. It represented aspects that were difficult to grasp or observe, but are still essential for life such as breath or wind. Based on alchemical practices, air was considered to be linked with the mind, intellect, and nonmaterial aspects of existence. It was the element of communication and movement. Water was the third element and symbolized fluidity, emotion, and intuition. It represented the flowing aspects of life and matter. Water takes the shape of what contains it which in alchemical terms translates to both adaptability and transformation. It was also associated with deeper aspects of the psyche and the realm of emotions. Fire was the fourth element and represented transformation, energy, and passion. Based on alchemical traditions, it was the force that creates and destroys which symbolized processes of change such as the combustion or digestion of food. Fire was seen as the element of purification and has been associated with creativity and the spirit. The four elements were not seen as physical substances, but they were also seen as spiritual and philosophical principles which described the different states of matter and the experiences human beings go through. Alchemists believed that we should understand and manipulate these elements that they were able to control the natural world and achieve personal transformation and enlightenment.

Seven planetary metals were linked to the seven classical planets in astrology. Each of them represented different physical and spiritual properties. Gold was associated with the sun and symbolized perfection, enlightenment, purity, and immortality. It was considered to be the most noble and the perfect of the metals which represented the ultimate goal of alchemical process.

Silver corresponded to the moon and represented purity, clarity, and reflection. It was associated with the feminine principle and the unconscious mind which reflects qualities such as intuition and mystery. Mercury was associated with the planet mercury. Based on alchemy, mercury was seen as the mediator between the spiritual and material realms. Copper was linked to Venus and symbolized love, balance, beauty, and artistic creativity. It was related to feminine aspects and was seen as a conductor of energy and the balance between the opposites. Iron was associated with Mars and represented strength, assertiveness, and resilience. It was connected to masculine energy and symbolized conflict and action. Tin corresponded to the planet Jupiter and seen as a symbol of wisdom, expansion, and prosperity. It was associated with higher thinking philosophy and the pursuit to obtain knowledge. Lead was linked with Saturn and symbolized transformation. Although it was seen as the basis of the metals, lead associates with saturnian qualities such as discipline and structure. They make it crucial for the alchemical process of transformation and transmutation. The metals were not considered to be physical substances but also represent various states of being, spiritual principles, and paths to personal and also metaphysical transformation.

The mundane elements refer to the common substances and materials that alchemists worked with experiments and symbolical interpretations. They are mundane since they are ordinary or what we call Earthly in comparison to more spiritually and astrologically significant elements and metals. Vitriol was known as sulfuric acid. It was essential to break down substances. It was seen as a powerful transformative substance which was capable of purifying or the power to decompose different materials. Antimony was used in purifying processes. It symbolized the wild and untamed aspects of nature. It was used to purify other metals. Arsenic was known for poisonous properties. Both arsenic and alchemy symbolized danger and transformation. It was used in many processes and was associated with transmutation. The mundane elements had a dual role in alchemy which served as physical substances used in the laboratory and as symbols for deeper spiritual and philosophical concepts.

There were also symbols that represented philosophical concepts. Alchemical symbols for philosophical reasons represented the blending of mystical, spiritual, and material ideas. As an example, The Philosopher Stone was considered to be a legendary alchemical substance that had the power to turn base metals into gold or even silver. It was also used to offer rejuvenation and immortality. It took many centuries to find the philosopher's stone. It was the primary objective in alchemy. The effort to discover the philosopher's stone was called the magnum opus. However, no one succeeded to obtain the philosopher's stone. The legendary stone embodies the final and important goal of alchemy. It was not just about physical transformation, but it was also considered to be the transmutation of the human soul into the higher enlightened state. It was similar to achieve heavenly bliss or the profound understanding of the universe. Ouroboros symbolized the concept of eternity and a perpetual renewal of life. This represented a core idea based on alchemical philosophy that everything was always evolving but still remains fundamentally unchanged. The green lion represented raw nature, the untamed and unrefined aspects of the world. It also represented copper acetate which was an important substance in alchemy. The green lion was sometimes seen as a symbol devouring the sun which symbolized the powerful and destructive nature of certain kind of substances and the idea that raw natural materials should be used to achieve higher states of knowledge. The white eagle symbolized both sublimation and purification. The white eagle expresses the movement from a lower baser state to a higher more refined condition in terms of both physically in substances and also spiritually in the practitioner. The black crow or raven symbolized the stage when an alchemist heats a substance to break it down to basic components just like the breakdown of age and false beliefs in personal transformation. The black crow gives rise to the destruction that must occur before there can be creation. Alchemy practiced in medieval times evolved into modern scientific disciplines especially in the fields of chemistry and medicine, but we should not think that alchemy is entirely dead. It still is alive through its symbols found in many aspects of modern culture from art and literature to psychology and also spiritual practices. We are now going to head over to an unknown journey to the world of alchemy.

Chapter 2

AN UNKNOWN JOURNEY

Who exactly were the alchemists? Could they have been magicians coming up with elixirs and potions or could they have been occultist mystics that had access to secret knowledge which was passed down from ancient times? Others also questioned the idea if they could have been charlatans who tricked gullible people to believe they could bestow riches and immortality for a right price? None of these carried out through, but these descriptions had some truth. Alchemy was considered to be a type of proto-scientific system which was a goal to understand, transform, and also purify natural materials.

The story begins in Egypt when no one knew the details of Egyptian alchemy. Even the alchemical texts are still unknown and have been lost. Starting from Egypt the system end up spreading around the old world through China, India, The Middle East, and Europe. Alchemy was considered to be diverse that changed from location to location as well as alchemy changing over time. There were different alchemists that focused on different things. Chinese alchemy was tied to taoism and traditional Chinese medicine. Indian alchemy focused on medicinal alchemy. Alchemy in the west and middle east had different goals which we will later discuss. We know a little more about Western alchemy since the alchemists during this time period wrote down their thoughts, theories, and experiments in codes and creative writings. Even modern scholars were not sure what the alchemists were up to. Codes and symbols with alchemy and astrology. There were several different types of astrological symbols and ideas used in alchemical texts. Seven planets were thought to correspond to seven metals. The sun represented gold, the moon represented silver, iron represented mars, tin represented Jupiter, copper represented Venus, lead represented Saturn, and Mercury represented mercury.

Alchemists also held to the idea to the Aristotelian idea that there were four classical elements such as earth, fire, air, and water. They also added three more substances which were mercury, sulfur, and salt. They even thought that mercury, sulfur, and salt were combined to make compounds such as metals and the bodies of living things. These mystical thoughts contributed to viewing the alchemists as sorcerers or magicians. They used magical language to hide knowledge and to use both spiritual and supernatural connections.

Alchemy did have three main goals, and these were the transmutation of base metals such as lead into precious metals such as gold which was called chrysopoeia, the manufacture of the elixir of life to cure illnesses and death, and the creation of an alcoholism or universal solvent to dissolve metals such as gold. The goals eventually got tied to the creation of the philosopher's stone.

We may ask ourselves that how did alchemists think of these goals as being possible or perhaps did they come from nowhere. Alchemists were similar to metallurgists, doctors, and natural philosophers who tried to understand the world around them. Alchemists were just like pre-scientific people who relied on wisdom of the past and observation. They were always synthesizing their own observations and experiments. An example of this was the synthesis of the transmutation of lead into gold. It is similar to like metallurgists heating metal ore where the ore would change color and different compounds appeared in the process. However, we know that these chemical reactions had impurities in the raw ore, but the alchemists thought there was evidence of a transformation of the material into a real and natural phenomenon. Even in our modern world, we observe the transformation of things found in nature. They felt they could use their transformative aspects in nature to their own advantage. Chrysopoeia was also thought of as a process of salvation. Transforming lead into gold was similar to the purification of a sinner through god's grace.

Then comes along the idea where we wonder about the philosopher's stone. There were a variety of processes that would create the stone. In fact, there were different texts which gave different instructions how to prepare this philosopher's stone. They felt it was achieved by heating various elements together until the elements combined and would go through a number of color changes.

The material first turned black to white to yellow to red and then in theory would become the philosopher's stone. The stone was then thought to turn base metals to gold by heating it with base metal. They felt it could be ingested by making the stone into a powder and then dissolve the powder into wine that would make the elixir of life.

There were many alchemists including those in the late medieval period and during the renaissance period who were obsessed with creating the stone. It was this creation that was known as the magnum opus or great work which became the main goal for later alchemists. There were even legends that would come about where someone has successfully created the stone which kept alchemists busy to continue to find the stone.

It comes as a surprise that there were famous scientists and philosophers who were alchemists. Isaac Newton was one of the inventors in Calculus, but he was also an alchemist in his spare time. Tycho Brahe was a famous astronomer and also an alchemist, whereas Robert Boyle who known to be one of the fathers of chemistry was originally an alchemist.

Even though alchemy has failed to reached its goals, it was still important as a precursor to the sciences of both chemistry and medicine. This ancient field still made important discoveries and inventions. Chinese alchemists created gunpowder, whereas western alchemists discovered some chemical compounds including sulfuric acid. Alchemists have also made contributions to the fields of metallurgy, ceramics, glass making, and even medicine. It was not until the fifteenth century, alchemy started to have a not good reputation.

People like Chaucer and others considered alchemists to be liars and quacks. Both Pope John XXII and King Henry IV of England made alchemy to be not legal. But alchemy was revived again during the renaissance period. But then again during the eighteenth century after the scientific revolution alchemy did not have a good reputation as charlatanism. After that the field of alchemy started to decline and modern day alchemists started to move away from esoteric and mysterious circles. But remember there would be no fields of alchemy if there were not discoveries of ancient elements we still use today.

Chapter 3

ANCIENT ELEMENTS

The earliest use of carbon and the oldest existing sample of carbon is from 26,000 BC. Both charcoal and soot have been known to the earliest human beings with some of the oldest charcoal paintings that dated to about 28,000 years ago located in Gabarnmung in Australia. The earliest industrial use of charcoal was the reduction of copper, zinc, and tin ores for the manufacture of bronze by the Egyptians and Sumerians. Diamonds were known to exist as early as 2500 BC. An early use of copper goes back to 9000 BC and the oldest existing sample was found to be about 6000 BC in the Middle East where Asia Minor is located at. The ancient element copper may have been the first metal to be mined and crafted by human beings. It was originally obtained being a native metal and later from smelting ores. Early estimates of the discovery of copper suggested it to be around 9000 BC in the Middle East. It was an important material for human beings throughout the Chalcolithic and Bronze Ages. There actually have been found copper beads that dated from 6000 BC found in Caralhoyuk in Anatolia and at the archaeological site at Belovode on the Rudnik Mountain in Serbia. This is where the world's oldest dated copper smelting took place from 5000 BC.

The earliest use of lead was around 7000 BC. The oldest existing sample was around 3800 BC. The discovers were in Asia Minor, and the place of the oldest samples were found in Abydos in Egypt. It was believed that lead smelting began around at least nine thousand years ago. The oldest known artifact of lead was considered to be a statuette found at the temple of Osiris. It was a site of Abydos that dated around 3800 BC. The earliest use of gold was probably before 6000 BC. The oldest existing sample was probably found to be before 4000 BC. The place where some of the oldest samples of gold artifacts were found to be in Wadi Qana in the Levant.

The earliest use of silver was probably before 5000 BC, and the oldest existing sample was found to be around 4000 BC. The place of oldest samples were Asia Minor. These have been discovered in Asia Minor shortly after both copper and gold. The earliest use of iron was before 5000 BC, and the oldest existing sample of iron was around 4000 BC. The plan of the oldest samples were found in Egypt. Yet there is still scientific evidence that iron was known before 5000 BC. The oldest known iron objects that were used by humans were some beads of meteoric iron that were made in Egypt in 4000 BC. It was the discovery of smelting around 3000 BC which led to the beginning of the Iron Age around 1200 BC. It was during this time period iron was used for tools and weapons.

The earliest use of tin was around 3500 BC, and the oldest existing sample was found to be around 2000 BC. The discovers were in Asia Minor, and the place of the oldest samples was in Kestel. Tin was first smelted in combination with copper around 3500

BC to make bronze. This is when The Bronze Age took place. Kestel which is located in Southern Turkey was the site of an ancient Cassiterite mine used from 3250 to 1800 BC. The oldest artifacts were dated around 2000 BC. Similar to tin, the earliest use of antimony was 3000 BC, and the oldest existing sample was around 3000 BC. The discovers were the Sumerians, and the oldest samples were found to be in the Middle East. There was an artifact that was said to be part of a vase was made of pure antimony which dated to about 3000 BC. It was found at Telloh in Chaldea which is part of Iraq. Both Dioscorides and Pliny describe the accidental production of metallic antimony from stibnite. But they recognized the metal as lead. The isolation of antimony was described by the Muslim alchemist Jamir bin Hayman from 850 - 950. The earliest use of sulfur was found to be before 2000 BC, and the oldest existing sample was found to be about 3800 BC. The discovers and the place of the oldest samples were found to be in the Middle East. Sulfur was first used around at least 4,000 years ago. According to Ebers Papyrus, a sulfur ointment has been used in ancient Egypt to treat granular eyelids. The Ebers papyrus was written about 1550 BC, but it was believed to be copied from earlier sources of text. It was designated as one of two elements for which metals were composed in the sulfur-mercury theory of metals. It was first described in pseudo-Apollonius of Tyana's "Secret of Creation" and also in the works attributed to Jabir bin Hayman during the eighth and ninth centuries.

The earliest use of mercury was around 1500 BC, and the oldest existing sample was found to be around 1500 BC. The discovers were the Egyptians, and the place of the oldest samples were in Egypt. Cinnabar was the most common mineral form of mercury(II) sulfide. It was a pigment from prehistory which dated back as the ninth millennium BC in the Middle East. There were cinnabar deposits found in Turkey being exploited from about 8000 years ago which also contained minor amounts of mercury metal. It was found in Egyptian tombs that dated from 1500 BC. The earliest use of zinc was before 1000 BC, and the oldest existing sample was also around 1000 BC. The discovers were Indian metallurgists, and the place of the oldest samples was found to be the Indian subcontinent. Zinc was used as a component of brass since antiquity before 1000 BC. The true nature of zinc was not understood during ancient times. It was found that a fourth century BC vase from Taxila was made of brass with a zinc content of 34%. It was too high to be made by cementation which also provided strong evidence that metallic zinc to be known in India by the fourth century BC. Zinc smelting was carried out in China and India around 1300. It was also identified as a distinct metal in the Rasaratna Samuccaya during the fourteenth century.

The earliest use of platinum was from 600 BC - AD 200, and the oldest existing sample was also from 600 BC - AD 200. The discovers were pre-columbian South Americans, and the place of the oldest samples were South America. It was used by Pre-Columbian Americans which was near modern-day Esmeraldas in Ecuador to make artifacts of a white gold-platinum alloy. There was a small box from the burial of the Pharaoh Shepenupet II who died around 650 BC was found to be decorated with gold-platinum hieroglyphics. The Egyptians probably did not know that there was platinum in the gold. The earliest use of arsenic was 300 AD, and the oldest existing sample was 300 AD. The discovers were the Egyptians, and the place of the oldest samples was found to be in the Middle East. Metallic arsenic has been described by the Egyptian alchemist Zosimos. Purifying arsenic was also later described due to the works attributed to the Muslim alchemist Jabir bin Hayyan in 850 - 950. The earliest use and the oldest existing smoke of bismuth was found to be in the year 1500. The discovers were the European alchemists and also the Inca civilization. The place of the oldest samples were found to be in Europe and also in South America. The Incas used bismuth with both copper and tin in a bronze alloy for knives. There were miners in the age of alchemy that gave bismuth the name tectum argent or “silver being made” meaning silver is still in the process to be formed within the Earth. We will soon look at how these elements have been used in The Stone Age.

Chapter 4

ANCIENT TECHNOLOGY

The origins of chemistry go back to the Stone Age. Early humans picked up and used naturally occurring materials. Useful substances were made by simple procedures such as crushing, burning, and boiling.

More complicated processes such as producing glass was developed. When the Greeks began to organize empirical observations into science that took place between 600 and 500 BC, there was a considerable amount of body of chemical knowledge. The Egyptians were informed about medicine and anatomy, while the Mesopotamian's were concerned about mathematics including elementary algebra and astronomy. But they never organized the information into a philosophic system. Even before the Greeks, no one had any concept about the laws of nature.

The earliest chemical processes were either one-step reactions that were discovered almost inadvertently or were developed step by step over a long period of time with each stage resulting in something useful. Ever since The Stone Age, people used pigments to color houses, boats, statues, pictures, and then themselves.

A sense of aesthetics has been a factor in humanity's cultural development. Ever since the beginning, people were attracted to colored minerals such as green malachite, light blue turquoise, and dark blue lapis lazuli. They used them, hammered them, and crushed them. They even noticed the rocks colored their skin. The process was going from accidental coloring to deliberate coloring. Between 30,000 and 20,000 BC, people grinned beautiful minerals into powders and then used them as pigments that were either dry or dispersed in water or soil.

Pigments were also used for cave paintings in places such as France, Spain, and North Africa. They were used to cover bodies of participants in religious and magical rites, such as places like New Guinea. Pigments were also used as cosmetics, where people painted their faces and eyelids. In the early days of Egypt, the upper eyelid was painted black with powdered galena (lead sulfide) and the lower lid green by using powdered malachite. In Mesopotamia, eyelids were painted with stilbite which is antimony sulfide. It is red in some forms and yellow in others. The purpose was to protect against fly bites, but then it shifted from repelling insects to attracting other human beings. In Egypt, lips and cheeks were painted with red ochre. The people in Mesopotamia preferred yellow ochre. Women painted their nails red and also tinted the palms and soles with henna. In East Asia, natural red palms and soles were considered good luck. During early historic times, fabrics were dyed. Prehistoric people used dyes, but the cloth itself was so perishable that there was no sample of dyed fabric that can be traced to The Stone Age. There is still evidence of prehistoric rock painting in the Sahara showing people wearing colored garments, and ancient wall paintings in Anatolic (in modern Turkey) that shows dyed woven carpets. Dyes are considered to be organic chemicals that are usually of vegetable origin. They are applied by steeping the cloth or fabric into a dye solution. Perhaps the idea of dyeing came from when people noticed the stains left by walnuts or berries or other types of foods. The first dyes may have resulted from accidentally wetting cloth with spilled soup that was made by boiling berries or other plants in a clay pot. When first fabrics have been colored by an organic material, it would have been natural to see what would happen if other plants or seeds were crushed or boiled.

Blue dyes were extracted from both wood and indigo. Yellow dyes were extracted by pomegranate, safflower, and saffron. Red dye was extracted from madder and henna. Kermes is also a red dye made from animal material. In fact, there was a species of insect growing on kernes oak that was collected, dried, and crushed. The coloring material was then extracted with an oil. The Phoenicians inhabited a few towns on the Mediterranean coast of what is now Lebanon. They used a local species of mollusk, Murex brandavis, to make purple dye. They ended up extracting the active secretion from shellfish, and then they steeped garments or cloth in a boiling solution of the extract. Only minute amounts of material could have been obtained from each mollusk. Tyre was known as the center for the production of purple. It was known for its foul-odor, which emanated huge quantities of rotting shellfish. The extract was expensive. Only the rich could afford the color purple. The royal purple became the symbol of nobility. In the

Byzantine Empire, an emperor, who was also the sun of an emperor, was called porphyriogenatos which meant "born in the purple".

Except for the colors indigo and Tyian purple, ancient dyes have not been colorfast. They tended to run and wash out unless certain substances called mordants were used to fix the dye to the cloth. Alum, which is also called potassium aluminum sulfate, was imported for these. It was actually used as early as 1000 BC and was known to have international economic importance during medieval times. There were people who wanted to color more than their fabrics. The Phoenicians made hair dye from alum, oil of cedar, and the sap of the shrub Anthemis tinctoria. The resulting hair color was probably light black or red. They really attracted the attention of people who were brunette.

Oils and perfumed oils have been used to protect the skin from the Near Eastern Sun. Perfumes have been extracted from flower petals and fruits. This was carried out by spreading them on layers of purified animal fat which absorbed fat-soluble oils and fragrances. The perfumed fat was shaped into balls and then placed on the heads of participants at feasts and then allowed to melt and run down. Perfumers also came up with other techniques for obtaining perfume from fragrant materials. They squeezed aromatic oils out of leaves and fruits and pressed them in a cloth bag. Early perfume-distilling process was carried out by or under the supervision of women. There were many early chemical procedures and equipment that evolved from cooking. There were actually cuneiform tablets that dated from the reign of the Assyrian king Tumult Ninth I (1256 - 1209 BC) who mention the name "the perfumeress" and her feral assistant.

It was also known that fruits, juices, milk, and other organic liquids ferment naturally. There were a variety of fermented drinks and dairy products known very early. People made wine and bear even before historic times. Wine was considered to be favorite drink in the Middle East except Babylon where beer was performed. According to the Bible, Noah found his sons to be drunken stupor. Lot's daughter got him drunk to seduce them who impregnating them. Hannah who was the mother of The Prophet Samuel was mistakenly admonished by a high priest to put the wine away. Nebuchadnezzar II, the mighty Chaldean king, imported wines on a large scale. Sargon II of Assyria had wine collars built in the palace. Ashurbanipal (688 - 626 BC) who was the last great Assyrian emperor was an expert on wines. In Egypt, the tax inspector assessed the quality of wines. Both Egypt and Mesopotamia wines were put into jars and were placed at a origin and sealed with an official seal.

In Mesopotamia, brewing was known to be a home industry. Women were the brewers and beer vendors. During the Sumerian period ending about 2600 BC, women had a privileged position in society. They owned property, had dower and divorce rights, can make legal agreements and sign contracts, and could follow their choice of occupation. In both Syria and Mesopotamia, women worked as distillers, chemists, and alchemists persisting until at least the Hellenistic era which was between 320 BC to 80 BC. In the code of King Hammurabi, the Babylonian law giver, the woven beer vendors have been around not to sell beer of low strength nor at a high price and not to allow political conspiracies to take place. Even these were mentioned in the policies.

Dyes, pigments, perfumes, cosmetics, wines, and beer can be obtained from available materials. However, there are other materials that cannot be easily obtained. Materials such as soap, metals, glass, and plaster are end products of complicated processes where the final product is different from the starting material. By the third millennium BC, the Sumerians have been using solutions of soap, even though solid soaps were not made until early medieval times. Soap was produced by boiling plant and animal fats and oils with alkaline solutions. We make these solutions from NaOH or KOH. During ancient times, they were prepared by leeching the ashes of wood fires with water to make a dilute solution of potash (potassium carbonate) and soda (soda carbonate). The solution was boiled and the fat or oil was dropped in. Fat or grease was dissolved slowly and formed a dilute solution of soap. It was then used for cleansing or removing grease. There was a series of steps or procedures that evolved slowly, each step resulting in a product. The new product or process served as the basis for the next step.

A possible sequence is as follows. Stone Age people used sand or ashes to scrape grease off their hands. If they rinse away the ashes with water, they may have noted the water to feel slippery. The ash-water mixture felt slippery because it contained alkali which dissolved the outermost layer of the skin. Later on in history, the Sumerians used a slurry of ashes and water that removed grease from raw wool and cloth so it could be dyed. Most of the dyes don't adhere to greasy cloth. Sumarian priests and temple attendants took the time to purify themselves before second rites. In the absence of soap, they used ashes and water. The more ashes there were in the water, the more concentrated the alkali solution. The material cleaned itself even after the ashes were not present. Eventually some put two and two together and discarded the wet ashes.

The slippery solutions clean because the alkali reacts with grease on the object and converts it to soap. The soap dissolves the rest of the dirt and grease. The more grease and oil dissolved by the alkaline solution means that there is more soap and the better the mixture cleans. People notice this since they were slippery solutions until the solutions lost their potency. The Sumerians realize a little grease improved the performance of alkali proceeded to make soap solutions by boiling fats and oils in alkali before using it for cleaning. Directions for making different types of soap solutions have been formed on cuneiform tablets. The final step in making solid soap was around 800 AD in Gaul. It was probably in the town of Savona, when soap was separated from water by salting out or adding salt to the solution. It was also possible that the Mesopotamians used salting out by 3000 BC.

Various products of pyrotechnology such as metals, glass, ceramics, and plaster have been invented during ancient times. According to Greek legend, the Titan Prometheus stole the secret of fire and decided to give it to humanity. The Greeks realized that civilization sprang from fire. Without energy from combusting wood, there was no way we could have produced the pottery, bricks, lime, weapons, and tools that helped us move out of The

Stone Age. In fact, pottery was the first product of pyrotechnology. The sequence of discovery probably went as follows. Primitive humans placed cooking fires on ground, but then they used cooking pots and holes in the ground lined with firewood. If the pit had been dug in mud or clay, the walls of the pit turned to stone rather than brick. We can think of clay as water softened stone or stone having water. When water was driven out by heat, the clay changes back to stone that is hard. The change takes place in the pottery kiln when ceramics become produced. Clay exposed to heat of a cooling pit becomes rock-hard. If a lump of clay was dropped into a roaring blaze, it should harden as well.

Once people notice the lumps hardened, they began to shape the soft clay in human and animal figures, heat them, and form ceramic figurines.

Baked clay dates back 20,000 years. Experimenters plastered gourds and wider baskets with clay and baked them to make pots, bowls, and jars. After 6500 BC, potters dispensed with gourds and baskets and shaped pots, bowls, and jars from wet clay.

There was more than pottery that came from fire pits. They wanted to get greater heat so potters tried different sizes and shapes of the fire pit. As early as 6000 BC, they transformed the fire pit into a furnace that reached temperatures high enough to melt copper. By 5000 to 4000 BC, artisans produced furnace temperatures high enough to transform copper ore into metallic copper. More trial-and-error improvements took place. Other metals were produced such as lime, plaster, and glass. An advantage of metals is plasticity. Metals can be stretched, bent, and melted. Metal objects can be easily shaped and reshaped. Wood has been known easier to cut or bend than metal, but it then wears out quickly and is fragile. Stone is more durable than wood and harder than either wood or metal. But stone is brittle and difficult to shape. Metals have a variety of uses that are not suitable for stone and wood.

There were high civilizations like the Egyptians, Mesopotamians, Chinese, and Greek who were dependent on metals. The Egyptians and the Greeks used iron tools for mining and also copper saws for cutting stone. The Palestinians and the Greeks decided to farm with iron plow shares. The Greeks trimmed and smoothed stone with iron chisels. They had metal clasps and pins used for clothing, metal hoops for barrels, and metal hinges and other furniture fittings. They also used metal clamps, spikes, pins, and dowels for stone columns. They also used structural iron beams in their temples. They had metal hammers, spikes, drills, files, rasps, knives, nails, arms and armor, and screws. The Egyptians and Mesopotamians used many of the implements of metals for wide uses.

But remember that ancient civilizations were not dependent on metals. Metals appeared in quantity during The Neolithic Era. Their uses developed gradually but it was not until the spread of the use of iron for tools and weapons about the middle of the first millennia BC did people begin to lose their dependence on stone, bone, and wood. Until 1000 BC, most Egyptian and Mesopotamian peasants almost had no contact with metals except for ornaments in temples, jewelry, arms, and armor seen on nobles and high-ranking officials. People also used wooden shovels, rakes, wheels, and mill wheels until after The Industrial Revolution especially in areas that were undeveloped.

The first metals known were copper and scarce quantities of gold and silver. Under ordinary conditions, most of the metals react rapidly with oxygen in air, so tin, mercury, iron, lead, zinc, and other metals are found in combinations with oxygen and/or other nonmetals. Gold and copper were discovered about 9000 to 7000 BC in the form of nuggets or lumps that looked like stones. Neolithic toolmakers picked up these stones and attempted to form them into arrowheads or knives. The stones did not chip, split, or flake normally how stones did. They heat and changed shape like clay. There was nothing in Neolithic experience that had prepared them for rocks that bent. It seemed magical. Here we can see an association of metals with magic.

The tool makers also played with fascinating pebbles by doing different things to them to see what happens. They found that after they were bent, they could be straightened out again and then bent into different shapes. They could also be beaten into flat plates and then cut into patterns. The new forms were ornamental and thought to be magical and were also in great demand such as gold because it never loses its beautiful luster. Whether gold ornaments showed status on the owner or were appropriated by those who already had status, chiefs and shamans wore gold, silver, and copper jewelry. The metals rapidly became associated with rank and power. It was easy to find and use metals, but it is difficult to produce them. There has been thousand of years that have passed between the discovery of natural metals and first smelted metals from ores. The relationship of metals and ores was not obvious. Even though minerals have been long used for pigments, no one could think of the connection between minerals and shiny metals.

Copper was the first metal produced from its ore since it was the easiest to smelt. The ore may have been malachite which was hydrated copper carbonate. When malachite was heated with either wood or charcoal at a high temperature, the carbon and carbon monoxide from the wood removes the oxygen from the ore and reduces it to metallic copper. The first smelting may have taken place in a cooking fire, but the chances that cooking temperatures were too low and oxygen in the atmosphere would have prevented reduction. Smelting requires high temperatures such as those taking place in a pottery kiln, so a kiln is probably where the first metal was produced. Perhaps a greenish copper pigment was used to decorate a clay pot before firing and the potter found a film of copper on the pot afterward. Perhaps a lump of malachite was used to prop up a pot. Someone may have put some rock or powdered mineral into a furnace or kiln, then the blue or green rock had disappeared and red copper was then present in the furnace. The rock turned to copper. The experiment was done over and over again. Other rocks were put into the furnace to see what would happen. Other metals and materials were produced. Copper, tin, and bronze were produced in useful quantities by 4000 to 3000 BC, smelted iron by about 1200 BC, and finally zinc in the medieval era even though small amounts of zinc have been made in Roman times. In China, iron ores contain large amounts of phosphorus and also melt at lower temperatures. Iron appeared in China much earlier than in the Middle East.

Glass was another product that came out of the pottery furnace. It was known as the world's first synthetic thermoplastic. Molten glass can be poured into almost any shape and then return the shape being cooled. Glass is a liquid at room temperature. It is too viscous to flow except except slowly and under pressure. It is transparent unless it contained air bubbles or undissolved particles. Ancient glass contained undissolved material and was probably not as transparent as ours. Glass was first produced as a byproduct of metallurgy. When it comes to mining, rocks can be dug up along with their ore, and when the ore is heated, the unwanted rock is heated to. The beginning steps in smelting convert the ore to metal but then leaves the rock unchanged. The result is a mixture of rock and particles or lumps of metals.

The earliest smiths broke the mixture and picked up the metal piece by piece. As years pass, improved furnaces attained temperatures hot enough to melt the metal and rock to form white-hot liquids that were insoluble to each other just like oil and water. In order to separate them, all the smiths had to do was pour the two liquids into a receptacle. The separation process was called liquation.

The early smiths found that after liquation when the molten rock cooled down, it formed a new material, a rigid, glassy solid, or slag. Heating all kinds of rocks and purifying their starting materials before smelting, they produced slags that were blue, green, brown, or even red and some of them were colorless and transparent. Actually they had made glass and they used the glass for beads and other ornaments, figurines, and flasks and beakers. There were other materials that could be put into the furnace to see what could be made from them. As early as 4000 BC, quicklime was produced from limestone by removing carbon dioxide. Quicklime was used to remove fat and hair from animal hides for the manufacture of leather. About 300 BC, quicklime's major function was used to make cement for construction. Plaster of Paris was also produced by heating.

The flames and searing heat of the furnace were dramatic chemical changes taking place seemed magical. Rocks turned to metal and glass. People thought this was the work of gods and demons. Gods and priests of metallurgy took place. People had their own god on the forge. He was Hephaistos to the Greek, Vulcan to the Romans, Well and Smith to the Saxons, and Try the One-Armed (whose name we got from Tuesday) to the Norse. Everywhere rites, sacrifices, and incantations have been developed to keep gods in good humor or to appease the demons. The rites had practical functions as well. They must have been effective in ensuring correct heating times and furnace conditions that had to be carefully controlled. For the smiths during ancient times, there was no way to measure temperature, tell time, or determine the composition of the ore. Control was not possible. Things often did not work. When something went wrong, people thought the gods must have been offended. They even offered a propitiating ceremony. When things did go right, it was important the illiterate smith remember what he did so he could repeat it next time. The ritual chants and incantations must also have been excellent mnemonic devices to help the apprentice learn correct procedures and timing.

From the earlier beginnings, metallurgy and pyrotechnology were connected with magic and ritual. This connection carried over into esoteric or religious alchemy. Technological information was kept secret. In both Assyria and Egypt, there were written recipes and lists of materials. As early as 1700 BC, papyri and cuneiform tablets contained warnings to the reader not to divulge secrets along with cursing on those who did. The Assyrians were considered to be enthusiastic antiquarians and got most of their information from ancient Sumerian texts. Some of them was simple and straightforward but many of them used a cryptic language that was intended to be understood only by the initiates. Copper was sometimes called eagle and ferrous sulfate to be a green lion. The ancient people had no concept of natural law, but they did understand cause and effect. They actually lived in a world where nature responded to ghosts, gods, and demons. It rained because the local rain god was either pleased or angry. Rocks and minerals were thought to be alive. They existed in male and female forms. They were born, grew in the ground, died, and had magical powers. To the Sumarians, each metal was associated with a god and a planet. They believed there was a strong connection between chemical and metallurgical processes and the stars. This belief persisted until the seventeenth century AD. Meteoric iron which came from the sky was considered to be sacred. Until the Greek philosophers time, technological phenomena were either religious or magical explanations. There was not even an attempt by the ancients to explain natural phenomena on a philosophical or scientific basis. This was odd since the Babylonians had accurate astronomical data and can predict eclipses.

In ancient Egypt, temples were centers of learning and intellectual activity. It was similar to monasteries in medieval Europe. Education was not confined to the priesthood. Many nobles and officers in the secular bureaucracy were literate, just like physicians and there were literate slaves to tutor children of the wealthy. The temples maintained libraries and schools for scribes, masons, jewelers, and others needed in the service of the temple gods. Many craftsmen have been engaged in the private sector of the economy. Imitation gold, silver, lapis lazuli, and other gemstones have always been in demand. The cost of real gold, silver, and gems was too much except for the royal family, the great nobles, and the higher echelons of priesthood. The bulk of Egyptian technology, astronomy, and mathematics were for the temples. Egyptian chemical processes have been permeated with magic and religion.

The situations were similar in Mesopotamia. In Egypt, the temples were centers of learning and knowledge. Stars and planets were observed nightly to keep track of the calendar of religious events. The god or goddess through the temple priests directed irrigation and flood control works as well as to the deity extensive tithes and voluntary offerings being bought. Priests needed to keep accurate records for large quantities of goods brought to the temple and distributed to workers and priests and assignments of work groups and performance. Writing and mathematics began with temple accounts but were needed for secular matters. Raw materials such as stone, timber, and metals were scarce in Sumar. People had to trade to survive. The Mesopotamians economy had a large private sector with shopkeepers, traders, merchants, and independent artisans. Merchants used to be literate and handle rudimentary arithmetic. There were secular scribal schools that trained them and their sons. Education was mainly practical and limited to matters useful to business. There was extensive written literature including romances, poetry, drinking, and love songs.

Since ancient Egyptian and Mesopotamian centers of learning were temples, the concept of nonmagical explanation of eclipses, earthquakes, pyrotechnology, and other phenomena was unthinkable and sacrilegious. A priest who has been trained to believe certain rites propitiated the god of glassmaking and who earned his living from the fears and offerings collected from the glassmakers cannot be expected to think about a chemical explanation for the process. If anyone offered such an explanation, he was considered to be a threat to both the god and the priest.

Egypt was an isolated but also a stable society. Maybe the adaptable Mesopotamians could have freed themselves from domination of their temples and theocracies and also developed the concept of natural law. But they did not have the opportunity. After the middle of the second millennium BC, they were a battered civilization, constantly at war, under attack by foreign invaders, under foreign domination and subjected to sack, pillage, massacre, and wholesale transfer of population. They sticked to the safety of traditions, and they did not embark on new philosophical ventures. Mesopotamian and Egyptian knowledge was transformed to people more favorably, the Minions and the Greeks, and from the populations new developments would soon to come. Natural philosophy awaited for the Greeks.

Chapter 5

CHEMICAL ANTIQUITY

It does not appear that differentiating any chemical material is essential to the existence of human beings. Distinguishable materials such as charcoal and ammonium salt came into existence at a early stage of civilization, but they may have been preceded by certain pigments such as the natural oxides of iron and manganese used by artists of the Paleolithic cave paintings at Altamira, Lascaux, and other places around 12,000 to 8,000 BC. Salts such as natron seems to have been preceded by common salt as an important entity in the Nile Region. Samples have been dated as early as the fifth millennium BC. Natron has been obtained from dry lakes, and it is primarily sodium carbonate or soda, adulterated with chloride and sulphate. It was used together with salt and gypsum for embalming, cleansing, and food preservation.

Archeologists have found evidences of metallurgy, pottery, and artifacts of glass-like composition. The techniques were practiced in Egypt and Babylonia and other places by 2500. Fired but not glazed pottery and brick began to replace sun-dried clay before 3000. The addition of natural mortars, clay and bitumen, artificial mortars, burned gypsum or Plaster of Paris and limestone (quicklime) have probably occurred at the same time. Glass has been found earlier than 2500. We think of it as an endpoint that began before 4000 BC. We find there is an increasing order of complexity in glass-like substances from the simplest glazed stones (quartz or steatite) in which stone was painted with soda and/ or lime water and then heated. Yet there are other more complex glass such as faience which is a moulded product that resulted from the heating of powdered quartz held together by soda or lime water. True glass which is about 1500 years later than glazed stones is a fusion of silica usually with sand that has 2-10% lime and 15-20% soda or natron. As mentioned in the previous chapter, this was not the end of ancient technology. Glazed pottery involved applying a glass-like surface to a porous fined clay which appeared 1,000 years after glass. A soda-lime-sand fusion product called grit also appeared during this time period. Ground frit was used as a blue pigment in Egypt about 2500 and has a definite composition of $CaO\text{-}CuO_4\text{-}SiO_2$. It seems to have been known as the oldest pigment. Another synthetic pigment this complex known in antiquity was the yellow lead antimoniate glaze found in Assyria from around 800. The pigment known as "Naples Yellow" is made by a different process. It was thought to have been found by heating altogether lead oxide or carbonate (litharge or white lead) and antimony oxide. These were products of the oxidation of minerals. By the seventh century, we encountered evidence of the use of antimony (sulphide) for decolorizing glass. Glass-like substances found in Babylonia were considered to be the site of technological innovation. Egypt and Persia were other similar sites of technological innovation.

Yet the chemical inventions interrelated to create the transformation of matter was in the field of metallurgy. Major production of metals except aluminum are still the basic metals of commerce which may have begun as early as 4300 BC. It was also completed with the introduction of brass about 4,000 years later. It involved chemical transformations through heating of certain minerals in a reducing atmosphere obtained by contact with fuel. The details of extracting metals from one another is different. Oxidation, reduction, and complex processes are each carried out to extract metals from ores. The minerals from which metals were obtained, the hydrated carbonates of copper (azurite and malachite), oxides of iron and tin (ochre hematite and cassiterite) and sulfides of silver and lead (galena) seem with apparent exception of tin oxide to have been in use as pigments prior to exploitation as metallic ores. Oxides of iron were used as pigments in prehistoric cave paintings. Other pigments found were malachite, galena, sulfides of arsenic, sandaracs or orpiment and stibnite.

The question of antiquity of artificially produced metals remained uncertain. It was not easy to differentiate nature from artificially produced metals. The discovery of artifacts Near Eastern archeology indicates that gold, electrum (naturally occurring alloy of gold and silver) and copper were extensive use of metals rather than precious stones before 4000 BC. By 3000 BC, this same condition existed with lead, silver, and bronze (coppertin) alloy. During the next two millennia both iron and tin joined metals of commerce, and later on copper-zinc alloy or brass came into existence. Most of these metals exist in the natural

state. Only gold and electrum have been derived from reactive metals in quantities by extant remains. Copper was the first metal artificially produced. The principle Egyptian copper ore (malachite) deposit at Sinai shows evidence as early as the Bahamian period. This was an era of three centuries beginning between 4200 and 3800 BC. There have been a few artifacts of copper and lead found in northern Mesopotamia from this era. Fabricated copper has been found in Iran from the preceding age. It was from western Iran which the Sumarian inhabitants of Mesopotamia have been emigrated. It was suggested they brought metallurgy with them. Efforts have been carried out to reconstruct the experiment that resulted from the reduction of malachite or azurite to copper. It was necessary to heat the ore, but to only heat it to a temperature attainable with an airblast and in a reducing atmosphere. Copper smelters, designed to operate with a natural draught, have been found late in antiquity. The origin of copper metallurgy have came about from the pottery kiln. This has been supported by excavations of pottery kilns from the middle fifth millennium. The transformation of the operation of copper being melted where its ore was smelled in a reducing atmosphere may have taken place between 4500 and 3500 probably in Iran but also in separated localities.

The priority of copper over lead came about as the evidence of excavations. This has been supported by those made in Egypt where large-scale workings of galena deposits does not appear before 2000 BC. This was about two millennia later from evidences of copper ore. Few artifacts of lead have been found in Mesopotamian excavations of about 4000. This justifies that galena could have been smelted in the neolithic cooking fire through heating at a low temperature and without a reducing agent for partially roasted galena acts that a portion reacts on the remainder to yield a metal. Artificial lead may have been preceded by copper; however, it was brought to our evidence that silver for which galena was the principle ore has been rare after 3000.

The second smelted metal appeared to have been was the copper-tin alloy or bronze, but this was still uncertain. Cassiterite was the only ore of one of the major metals which has not appeared to have been used as a pigment nor do we find it in Egypt or Babylonia. There were complex ores that contain copper and tin that did exist. But they were rare and have failed to produce bronze in modern experiments. Cassiterite occurs as an unusually heavy stream pebble, and it was noticed in panning for gold. It somehow found its way into the copper smeltery. These accidents occurred before 3000 BC and came from the same Iranian region where the discovery of copper metallurgy took place. Tin can be smelted at a lower temperature but unlike galena the ore required the addition of a reducing agent. This is also true of rarer metals such as antimony and bismuth, but they are easily confused throughout antiquity. Iron was the last of the classical metals to enter commerce in smelted form. It required equally rigorous conditions similar to copper smelting. It was still scarcely metallic without further complex treatment. During this time period, iron was a rare metal and precious.

European travelers had been bringing home specimens of Egyptian handiwork since the seventeenth-century. There was a group of French scientists who accompanied Napoleon on his adventures to Egypt in 1798. This was probably the first systematic archeological expedition. The famous Rosetta Stone was found and a quarter century later Champollion began the decipherment of Egyptian writing which made numerous papyri and stone inscriptions useful. Expeditions were sponsored by France and Tuscany in 1827 and by Prussia in 1843-1845 after which systematic archeological exploration of Egypt to be continuous. Museums in Europe and America gathered great quantities of literature and artifacts of ancient Egyptians. In Egypt, the government established a museum in 1858 that now maintains the largest and most important collection of Egyptian antiquities.

In western Asia, the transition from unsystematic collection to systematic excavation took place in 1842-1845. This was an area identified as the ancient site of Assyria. The decipherment of uniform writing on clay tablets was found by 1846. But only in recent years have metals, pigments, cosmetics from near Eastern excavations was dated and analyzed. From Western Asia, we have a medieval tablet from Nippur that was dated 2100 BC and a tablet of seventeenth century BC relating to glass coloration. Information on practical chemistry occurred in the remains of "Library of Ashurbanipal" (Assyria) and dated in the eighth and ninth centuries BC. In Egypt, we find a number of papyri relating to medicine. The papyrus Ebers mentions a considerable amount of drugs. The oldest papyrus relating to chemistry appeared to be a set of jewelers recipes in the Leyden and Stockholm papyri from the third century

AD. These date from the nineteenth century excavations with the exception of the tablet on glass coloration.

The Babylonian glass text of the seventeenth century mentions that lead, copper, lime vinegar, copper acetate, and saltpetre. It is interesting to learn that saltpetre was also found in the Nippur medical tablet of about 2100 BC. There were other materials found during this time period even though we lack evidence of and these include sal ammoniac, brass, and fuming sulphuric acid. Some of the materials known in remote antiquity has been lost. The Egyptian document, which was the medical papyrus of Ebers of the sixteenth century BC does not mention any of these materials, but it does list the practice of chemistry as we know from archeology.

Materials such as glaze or glass, gypsum, asphalt, and charcoal in medical texts was common during the next three millennia. Drugs have been considered to be reliable guides to the total of inorganic materials. It was not without justification that a brief list of minerals in the pharmacopoeia of the Greek, Dioscorides, was regarded as alchemists as a compilation of important substances.

Chemistry has changed very little during the twelve centuries between the papyrus Ebers and the oldest extant Greek treatise on the practice of chemistry. The development of thermo-chemical processes which includes metallurgy, the production of artificial minerals, and the development of a process for obtaining gold were deduced from the Babylonian-Egyptian sources. There were other landmarks that occurred in GrecoRoman antiquity.

With the advent of the millennial domination of Greek literature starting from 500 BC was given a new foundation. Greek natural philosophers established themselves during this time period. Evidence of practical chemistry appeared in Greek literature as early as Homer and an early natural philosophy reveals a familiarity of materials. But still the earliest surviving Greek treatise that deals with the subject was the short work called "On Stones" written about 300 BC. By Theophrastus who was Aristotle's pupil and also the successor as head of Lyceum at Athens. Aristotle wrote extensively on natural philosophy and begun the tradition of natural history with an encyclopedic description of the animal kingdom. Theophrastus wrote on plant and mineral kingdoms and the latter work deserves to be called the first treats on practical chemistry.

Theophrastus was inspired with his descriptions of stones rather than by the methods of metallurgy, medicine or coloring but only the minerals used in the arts. He divides the substances found in Earth into metals such as gold and silver, stones, and earths. He was concerned with the precious stones and building stones and referred to his "earths" as "immature stones". He differentiated them based on properties such as the physical properties of color, hardness, frangibility, inflammability, and power of attraction (amber, magnet). He felt the "Earths" had special qualities and powers. This explains that they were chemical. The earths include with the exception of tin oxide are the pigments as the first ores of common metals. They also include gypsum, bitumen, cinnabar, and two arsenic sulfides such as orpiment and realgar and varieties of other clays. He also includes white lead and copper

acetate neither of them which occurs in nature. They have been prepared through the corrosion of metals by vinegar.

Encyclopedic description of the arts and sciences was known among the Romans, and Greco-Roman literature actually provides for the first time comparisons. It is simple for us to compare Theophrastus "On stones" to the Natural History of elder Pliny and Material medica of Dioscorides who were both from the first century AD. Even though the purposes were not the same, similar practical chemistry emerged from these three comparisons. This shows another indication of ancient practice chemistry taking place during this time period. A long history of mining and metallurgy was led by Theophrastus, Dioscorides, and Pliny to the introduction of a few new mineral materials. Open-pit mining and new ores were being exploited. Pliny's principle ore was silver sulphide ore (argentite) and his copper ore was the iron-copper pyrite, chalcopyrite found below surface deposits of malachite or azurite. It was also apparent that other minerals have been turned up in metal mines. Most of Theophrastus "earths" are in the vicinity of metal mines. However, Pliny was more interested in other things found with metallic ores than in the ores themselves. Based on silver mining, the other things were cinnabar and antimony sulphide (stibnite) and the principal other material found in copper mines was cadmic (zinc silicate or carbonate). The practice of metallurgy produced by-products like the oxides of metals.

It was still difficult to differentiate substances as the small mineral continued to go through several names based on geographical distribution of production. One of more oxides of a metal of lead were added to the alreadyknown compounds of natural occurrence. The yellow and red oxides of lead (PbO, litharge; Pb_3O_4, minium, red lead) were known to Pliny and Dioscorides. They associate them not with lead but with silver which appeared to have been the objective of the manipulation of galena. Discorides described two yellow substances molybdana and lithargium which appeared in the course of smelting gold and silver. He states that lithargium was obtained from lead sand. But we infer from the sources and color they were from litharge. Pliny gives us a clear account of the same substance of the scum of silver which also turns up during the course of smelting that metal. He also describes molybdana the same as Dioscorides, but then he confuses it with the sulphide (galena). The red oxide, minium, was known, but there was no indication that it was obtained by further oxidation of litharge or the two were even related. Dioscorides describes two substances ammion and sandarach and we infer both of them as minium. The former was obtained by heating white lead which was the product of prolonged action of vinegar on lead. Pliny describes two substances minium and sandarach. His description of the former was identical to Dioscorides description of ammion, but Pliny's sandarach was found in gold and silver mines. He even notes that a false sandarach was made by heating white lead. We can conclude that Pliny and Dioscorides have known the oxides of lead. The same argument can be said of the oxides of copper. They knew more about the oxides, but had a poor conception of what an oxide is than Plato. Plato had the idea that such substances to be the result of

weathering of an earthy part of a metal. Except zinc oxide, those that sublimate or decompose on strong heating such as the oxides of arsenic, antimony, and mercury don't appear in the ancient works of ancient writers. Zinc oxide appeared in quantities in the flues of brass furnaces. Others have been overlooked or confused with something else. Oxides of antimony found on the part of the Assyrian pigment makers produced "Naples Yellow".

Chlorides, sulphates, and the exception of vitriol-alum group sulphates have not been known since they occurred naturally. It was doubtful they were made artificially instead incidentally. Silver chloride appeared as a byproduct of the salt cupellation of gold. Pliny states this residue as wasted gold. Dioscorides procedure for burned copper recommends heating it by itself in air or with salt, sulphur, or vinegar which might yield the chloride, sulphide, and acetate.

There was less opportunity of the acetates, especially those of copper and lead were made early. Lead acetate was not differentiated from the pigment white lead, and copper acetate (verdigris) was the only product of vinegar dissolution that was clearly discriminated. Even though the vitriols and alums were known as classes of substances that differed in their metallic constituents, iron vitriol (ferrous sulphate) and potassium alum (potassium aluminum sulphate) were commonplace materials passing under the names vitriol and alum. The metallic elements were among the common in the earth's crust. The combination of sulphur was less common, but occurred in regions of one-time volcanic activity. The rarity of vitriols and alums was

attributable to their solubility. Where rain was common, they may have observed efflorescences on rocks and under certain conditions form crystals. Pliny called these crystals "this kind of salty earth".

Alum has been traced from two of its principal uses which includes it being an astringent medicine and fixing dyestuffs near the beginning of the second millennium BC. There has been evidence of its production claimed thousand of years even before that. Iron vitriol was associated as an impurity and has an astringent taste. It was suspected that alum of antiquity was vitriol, but the dye fixing property does not belong to vitriol. This showed that alum was known to be found during antiquity.

The "salty earth" of Pliny which he calls "alumen" comes in several colors and was produced from Armenia to Spain.

Every type comes from water which exudes from the earth, but it was also formed by cooking the stone chalcitis. Astringency would be the Greek name and stypteria would be its property but in liquid form is also indurative which is susceptible to crystallization and corrosive. It was also used in dyeing as a mordant, in the field of medicine, purifying gold, and blackening leather. The first use refers to alum, the last two to iron vitriol, and the name chalcitis to copper vitriol. The less common blue vitriol should enter the picture since it occured in the drainage of copper mines. The Romans were interested in this, and it was claimed the Romans thought copper to be the essential constituent of alums and vitriols. But they were differentiated by those who use them and there was evidence that they were produced in large quantities.

It should not be surprising that the subtle differences between common soluble white salts, sea salt, the alkali carbonates, sulphates and nitrates were recognized. Salt which meant sea salt stated by Pliny to be of two kinds with several varieties. The two kinds are natural and prepared and includes some varieties identical with his natron. Natron accrued some varieties of its own. Both Theophrastus and Pliny mention a variety was obtained by leaching plant ashes. This could have been potash (potassium carbonate) from ordinary plants or soda (sodium carbonate) from different plants near the sea. During the Middle Ages, these sources of alkali were familiar to Europeans, and the product was purer and more homogeneous than Egyptian natron. It was then the unique properties of the alkali carbonates became known.

Colorless salts such as the nitrogencontaining salts sal ammoniac and saltpetre were difficult to distinguish. These could be called the salts of civilization since they were obtained from the refuse of domestic animals. Each of the two salts had a revolutionary effect on chemistry when it entered the practical chemist in the ninth century (sal ammoniac) and the thirteenth century (saltpetre). The conditions for their formation existed in populated regions of the ancient world observed. Pliny records in his discussion of colorless salts such as the formation of sodium hydroxide (caustic soda, lye) by treating natron with quicklime. This was the first substance known capable of spontaneous chemical activity. Caustic soda was used for the manufacture of soap. We don't know if Pliny knew this was true soap, but it was known to Galen during the third century AD. It was common in Europe by 800

AD. It may have been the discovery of the Gauls who were far from the mainstream of chemistry. Pliny did not consider its properties remarkable, but its conspicuous property, causticity, was known in quicklime from the sixteenth century BC. Pliny regarded caustic processes of natron with quicklime as something intense.

Two metallurgical innovations came about by the Greco-Roman writers were the introduction of mercury (quicksilver) and brass. There were many isolated samples of metals found belonging to early periods and this is the case with mercury and brass. It still does not appear that mercury was known before that date about 400 BC which Theophrastus gives the invention for a process of production by rubbing cinnabar in vinegar. It appears in chemical treatises that making it was simply by heating cinnabar. The importance of mercury during the late antique period and medieval chemistry was still not emphasized.

Earliest known examples of brass predates the appearance in excavations. It was held by some that brass was made in Persia from about 700 BC. They are finer ground regarding the Romans as the first fabricators. Pliny included brass among coppers improved by smelting with a substance which was found in silver mines (calamine or smithsonite) and appeared as a sublimate (zinc oxide) in the flues of copper furnaces. Both of them were usually called tutia. During this time period, brass was not recognized as an alloy, but it was regarded as another type of copper. It was also difficult to differentiate copper from its alloys. Just like caustic soda, it was regarded as an improvement of a substance already known. It also had little chemical significance. As for the metal zinc, it still remained unknown until the sixteenth century AD.

Surviving papyri and cuneiform tablets give a limited hint of theoretical basis of practical chemistry. The existence of any chemical theory before the Greeks has been questionable. It was likely the theory has been lost. Combining evidences from the practice of modern primitive tribes with inferences drawn from alchemy has been made. The animistic metallurgy seems plausible than extraordinary spectacle of primitive chemistry.

During primitive societies, minerals participate in a sacred character of earth. Just like animals and vegetables, minerals have life. After an embryonic growth in earth, they reach maturity in a state we know as metals. The miner and metallurgist intervene and accelerate normal maturation. They view the ore-reducing process as magico-religious relationship with nature. This was similar to an agriculturist who speeds up the growth of plants. But still the metallurgist was spectacular and mysterious. He was considered to be in primitive societies as the position as "master of fire" and a member of an occult religious society which are the secrets transmitted by rites of initiation.

Primitive theory shows little light on the evolution of the idea there was a group of substances that was limited in number and which possess the distinctive properties as metallic. This emerges through evidences of archaeology and early technical writing with literature of Greek natural philosophy which exhibits the workings of nature.

The Greeks were different in that they felt the term metal was denoted as a valuable mineral site then in Herodotus

450 BC, a mine and finally a produce of a mine. According to Lucretius during the first century AD, it was used in a modern sense. Among the Hebrews, Greeks, and Romans, a metal was in simple terms a valuable mineral and the values of metals was listed in an order which remained consistent such as gold, electrum, silver, lapis lazuli, malachite, copper, iron, and lead.

Another ancient criterion of classification was associating metals with planets. The development of association still gives us some clue to the differentiation of metals as distinct substances. Associating gold and silver with the sun and moon is prehistoric antiquity. Five planets were differentiated as astronomical objects and five metals were associated with them. Copper, iron, and lead were associated with the planets Venus, Mars, and Saturn. The planets Mercury and Jupiter were reasons the planets for which other metals competed. Evidence also shows that tin, bronze, electrum, and quicksilver which was known after 300 BC, alternating as the metals associated with Mercury and Jupiter. The uncertainty reflects an imprecise differentiation of principal alloys, electrum, bronze, and during Roman times brass from metals on which they were based, the imprecise differentiation of tin and other metals such as antimony and bismuth from lead, and the unique qualities of quicksilver.

Yet still these lists indicate that the definition of metals as a particular class of mineral substances was complete by the second century. Copper alloys persisted on these lists, and differentiating members within the class of metals was not complete until the end of the eighteenth century

AD. This was delayed by contradictory ideas that there was a fixed number of metals (about seven) and the metals differ by imperceptible degrees of perfection from the base lead to the most perfect gold. It was also held by the beliefs of Plato and Aristotle that it was more or less indiscriminately transmutable. Testing for purity was a matter of convenience. There was little evidence that the list of coinage metals was taken seriously. Discovering the method of separating gold from metals alloyed marked a milestone in the history of chemistry. According to Agatharchides during the second century BC, the process consisted of manipulating adulterated gold being heated with salt, straw, and other materials. The time of discovery was fixed by the disappearance of electrum from the Egyptian lists of metals after the seventh century BC. Yet electrum still survived on lists elsewhere. Agatharchides process as a test for gold was not a useful model of testing since it depended upon the resistance of gold to all chemical treatments that altered adulterants. Only silver and iron were defined by writers through the European Middle Ages to have characteristic properties and no other important alloys except electrum in the case of silver. During the same period, copper, bronze, and brass were confused and lead seemed to have been the chief member of a subclass that included tin, antimony, and maybe bismuth. It was once thought that metallurgy was not a lack of technique that inhabited the differentiation of metals and alloys, but a lack of awareness the metals differed distinctly and not by infinitesimal gradation. The difference of the seven metals seemed to have owed more to the field of astrology than chemistry.

The origins of chemistry have taken place during ancient times, and there still lacks some records and documents what was found during these time. The rise of Greek philosophy and civilization as well as matter and its changes took place many years ahead. Postsocratic philosophies and microphysics in antiquity helped continue to build the foundations of science. Much historical changes including the "Physic et Mystic" which was the scared text of Greek alchemy took place. The origins of chemistry was complex but eventually all of this lead to the meaning of alchemy.

Chapter 6

MEANINGS OF ALCHEMY

Alchemy was understood to be the art whose goal was to transmute base metals into gold using The Philosopher's Stone. Even from a standout, this was a superficial view. Alchemy at once was thought of as a philosophy and as an experimental science. It was the goal to transmute metals that would give the final proof of the alchemist hypothesis. Alchemy was considered to demonstrate on the material plane of a philosophical view of the Cosmos. One of the alchemists would state:

> ***"Would to God…all men might become adepts in our Art—for then gold, the great idol of mankind, would lose its value, and we should prize it only for its scientific teaching."***

However, not all alchemists agreed on this. Alchemy meant making gold and gaining wealth.

Some mystics expressed that Alchemy was not considered to be a physical art or science for the manufacture of gold and the process itself was not carried out on a physical plane. Based on transcendental theory, alchemy was more concerned about the soul, the object for perfection and not material substances, but in a spiritual sense. Those who held this belief felt alchemy was a branch of mysticism from a special language. They felt the writings of the alchemists was not just about dealing with chemical operations such as using furnaces, retorts, alembics, pelicans and with salt, sulphur, mercury, gold, and other substances but instead thought of as allegories dealing with spiritual truths. Transmuting base metals into gold symbolized salvation. It was the transmutation of one's soul into spiritual gold. This can only be obtained by eliminating evil and the development of good based on the grace of God. It was the realization of salvation or spiritual transmutation that may be described as the New Birth or the condition being known as union with the Divine. It follows that the alchemists were pure mystics.

The development of chemical science was not due to labor, but to pseudo-alchemists who misunderstood their writings that interpreted them.

However, this theory was not all accepted. The lives of the alchemists were indeed occupied with chemical operations on the physical plane. They did discover methods for transmuting common metals into gold. Paracelus calls the alchemists of this period as "Spagyric physicians".

The writings of the alchemists contain accurate accounts of chemical processes and discoveries which still cannot be explained by transcendental interpretation. Chemistry really owes its origins to the alchemists themselves and not to others who misread their writings.

It is still important to note that there was a considerable element of mysticism in alchemist doctrines. However, those who approached from the scientific point of view have considered the mystical element as little to no importance. There are facts that contain purely physical theory of alchemy, and the importance of this mystical element and the relation that existed between alchemy and mysticism was essential to understand the subject. The alchemists always speak of their Art as a Divine Gift which are considered to be the highest secrets not to be learned from any books on the subject, and they teach the right mental attitude with God as the first step for the achievement of the magnum opus.

In the second place, we should also notice the nature of the alchemistic language. The language of alchemy was highly mystical and there was much unintelligible in a physical sense. The alchemists apologize for the vagueness that mighty secrets may not be made manifest. It was true during the days of alchemy that there was a good deal of pseudo-mystical nonsense. It was written by imposters, but still the mystical language was by no means confined to later alchemistic writings. The alchemists shielded their secrets from vulgar and profane eyes and would adopt a symbolic language.

There was past belief that the language of alchemy was due to some plan. The argument cuts both ways especially for those who take a transcendental view of alchemy as its language as symbolical but in a different manner. The mystical element should be found in writings of earliest alchemists. The manuscripts have not been written for publication. It did not risk informing the vulgar of the precious secrets of alchemy. The transcendental method of translation succeeds in what was making sense and what was not intelligible in the writings of the alchemists.

Alchemists loved symbolism, and it was displayed in designs which certain books embellished. We are not referring to illustrations of apparatus employed in operations of physical alchemy, but instead to pictures whose meanings have reference to physical or spiritual processes. It was also known the majority of alchemists were involved in problems and experiments of physical nature. There were few of those concerned with problems of spiritual nature. Western alchemists make an appeal to Hermes Trismegistos as the greatest authority on the art of alchemy whose writings were undoubtedly a mystical character. It was clear that alchemy was apparent in physical nature and mysticism as well.

If we are going to understand alchemy, then we must look at it from an alchemist view. During modern times, there has come a divorce between religion and science, but it was different according to alchemists that their religion and science was closely united. We stated earlier that alchemy demonstrated experimentally on the material plane the validity of the philosophical view of the cosmos. We can see now this philosophical view of the cosmos was mysticism. Alchemy has its origin in an attempt to apply the principles of mysticism to things of the physical plane. It was of dual nature meaning that it was spiritual and religious, and on the other hand physical and material. Alchemy and mystical regeneration are similar processes on different planes because they are founded on the same first principles.

The famous axiom known by every alchemist:

> ***"What is above is as that which is below, and what is below is as that which is above."***

This is what we can express the basic idea of alchemy. The alchemists postulated and believed the essential unity of the cosmos. There is an analogy existing between spiritual and physical things which are the same laws operating in each realm.

The alchemists held that metals are one of essence and spring from the same womb of nature but it is not equally matured and perfect. Gold was held as the highest product of Nature's powers. In terms of gold, the alchemist envisioned a picture of regenerate man who is resplendent with spiritual beauty and overcoming all temptations and proof against evil. On the other hand, lead which was the basest of the metals was typical of the sinful and unregenerate man, stamped with hideousness of sin and overcome by temptation and evil. Gold withstood the action of fire and known to withstand corrosive liquids, while lead was most easily acted upon. The Philosopher's Stone was said to bring the desired grand transmutation, and was the species with gold and purer than the purest. It was understood in a mystical sense that the regeneration of man was effected by goodness itself. The Philosopher's Stone was considered to have incredible powers attributed to it.

The alchemists view was between the problem of the perfection of metals or the transmutation of base metals into gold and the perfection of spiritual man.

It might also be added between these problems of the perfection of man considered to be physiologically. The alchemistic philosopher considered these three problems as one. It was the same problem on different planes of being. The solution was one and the same. Those who held the key to one problem held the key to all three of the problems. This was based upon the analogy of matter and spirit. The point was not whether the problems were one and the same. The main doctrine of analogy was an essential element in all mystical philosophy meet with general consent. It will be contended that analogies drawn by alchemists were fantastic and were not always correct. There may be more truth in them that appears at first sight. The point is not that the analogies are correct, but they are regarded by all true alchemists. A considerable portion of alchemistic work was taken as an analogy believed to exist between The Philosopher's Stone and stones in general used for buildings during ancient times.

For the most part, alchemists were engaged with carrying out the alchemistic theory of the physical plane. It was the goal to transmute base metals into noble ones. It was the love for knowledge and for the love of mere wealth. Those who were worthy of the title "alchemist" realized at times the possibility of the application of the same methods to man and the result of transmutation of man's soul into spiritual gold. There were a few who had a clear vision of this ideal, those who devoted their activities to the highest goal of alchemistic philosophy, and those concerned with the analogous problem on the physical plane. The theory that alchemy originated to demonstrate the applicability of principles of mysticism to things of the physical realm brings into harmony both the physical and transcendental theories of alchemy and the conflicting facts that exist between them. It helps explain the existence of two types of alchemists. It explains the appeal to the great works of Hermes and the presence of the writings of the alchemists that was mystical. We will continue to look mainly to the physical aspect of alchemy. In order to understand its theories, it appears to be essential to realize that alchemy was an attempted application of the principles of mysticism to things of the physical world. The analogy between man and metals sheds light on what would be difficult to understand. It helps to explain why alchemists attributed moral qualities to the metals which some call them “imperfect” or “base”. Others were said to be “perfect” or “noble”. It does help explain the notions regarding the nature of the metals.

The alchemists believed that metals were constructed after man into whose constitution three factors were regarded as entering the body, soul, and spirit. Mystical philosophers use the word "body" as an outward manifestation and form, "soul" as an inward individual spirit, and "spirit" as the universal soul of all men. According to the alchemists in the metals, there is a "body" or what we call outward form and properties, "metalline soul" or spirit, and then finally the essence of all metals. Note that the alchemists have not always been consistent with the term "spirit". Sometimes they employed it to be denoted as the more volatile portions of a chemical substance, and there were times it had a more interior significance. There was a great difference between the alchemistic theory and views held in modern chemistry. We do find the similarity between alchemistic theory of a soul of

all metals (a one primal element) and modern views of science. In an attempt to demonstrate the applicability of the fundamental principles of mysticism to things of the physical realm of alchemy failed and ended its day and time in fraud. It appears that the true aim of alchemistic art was the demonstration of the validity of the theory that many forms of matter have been produced by an evolutionary process from some one primal element or quintessence in the domino of physical and chemical science.

It was believed that the transmutation of metals was a common occurrence. As an example, depositing copper on iron when immersed in a solution of copper salt or blue vitriol was concluded to be a transmutation of iron into copper. If the alchemists examined the residual liquid, they have found that two metals simply exchanged places.

White and yellow alloys of copper with arsenic and other substances could also be produced pointing to the possibility of transmuting copper into silver and gold. It was known that water was boiled for some time in a glass flask and then some solid earthy matter was produced. If there was a way water could be transmitted into earth, then one metal could be converted into another metal. The alchemists regarded the transmutation of metals as an experimentally proven fact. Even if they have to be blamed for their superficial observation, the labors marked a distinct advance of speculative and theoretical methods of the philosophers preceding them. Whether they were at fault, the alchemists were forerunners of modern experimental science.

It was interesting to note that the alchemists regarded metals as

composite. They felt the transmutation was a logical conclusion. If we are going to understand the theory of the elements, then we must step back away from the views of our modern world. It was still a fact of simple observation that different bodies manifest itself in properties such as combustion. Properties were regarded as some principle or element common to all bodies exhibiting the properties. Combustibility was thought to be due to some principle of combustion perhaps the sulphur of the alchemists and the “phlogiston” of the later period. This view which a priori appeared to be not likely. It was known there are relations that exist between properties of bodies and their constituent chemical elements and the arrangement of the particles of the elements. It was not obvious properties that enable chemists to determine the constitution of bodies and the connection far from being simple imagined by alchemists.

We talked earlier about the origins of practical chemistry in the preceding chapters. Let us revisit that picture again going back to the philosophers preceding the alchemists. It was not improbable that they derived it from older sources. Empedocles of Agrigent (440 BC circa) felt there were four elements such as earth, water, air, and fire. Aristotle decided to add a fifth element "the ether". The elements were regarded as not different types of matter, but rather different forms of one original matter where it manifested different properties. It was once thought that these elements were due to four primary properties of dryness, moistness, warmth, and coldness where each element being supposed to give rise to two of the properties such as dryness and warmth being thought to be due to fire, moistness and warmth

to air, moistness and coldness to water, and dryness and coldness to earth. Moist and cold bodies especially liquids were said to possess the properties in consequence of the aqueous element and was termed "waters". The elements were not regarded as different kinds of matter. Transmutation was still thought to be possible. Based on the alchemists, we find the view that metals were made up of two elementary principles such as sulphur and mercury in different proportions and also degrees of purity. The terms "sulphur" and "mercury" must not be understood the bodies by their names. Like the elements described by Aristotle, the alchemistic principles were found to be properties instead of as substances. It was still unclear by the alchemists themselves. It was not easy what the alchemists meant by these terms. They mention different sorts of "sulphur" and "mercury". They felt that "sulphur" was regarded as the principle of combustion and color and was present that most metals were changed into earthy substances with the help of fire. As for "mercury", the metallic principle "par excellence" had to do with properties as fusibility, malleability, and lustre. These were characteristics of the metals. One alchemist felt the excess of sulphur to be a cause of imperfection in the metals and there was corruption of the metals of fire. He also made the assumption that metals contained an incombustible and a combustible sulphur where the latter sulphur regarded as an impurity. Another alchemist felt that sulphur was recognized by the vital spirit in animals, the color of metals, and odors of plants. They even felt mercury was the cause of perfection in the metals and endows gold with its lustre. There was even another alchemist that felt quicksilver was the elementary form of all things fusible, and when these things fusible melted they changed into it and mingles with them the same substance with them. The bodies differ from quicksilver and was not free from foreign matter of sulphur. It was the obtaining of "philosophical mercury" that was held to be essential for the attainment of the magnum opus. It was thought it could be prepared from quicksilver by processes of purification where the impure sulphur that was present might be purged away.

The sulphur-mercury theory of metals was held by famous alchemists Roger Bacon, Arnold de Villanova, and Raymond Lully. It was thought to have originated from the Arabian alchemist Geber. Late Professor Berthelot showed works ascribed to Geber in which the theory were forgeries of a date which are centuries old. Arsenic was regarded as an elementary principle, but this idea was not general.

As time progressed, the mercury-sulphur theory was extended by adding a third elementary principle called salt. In the case of philosophical sulphur and mercury, the term was not meant common salt such as sodium chloride or any of the substances known as salts. “Salt” was given to the name of the basic principle in the metals having the principle of fixity and solidification. It was also understood by conferring the property of resistance to fire. The theory was found to be in the works of Isaac of Holland and Basil Valentine to explain differences in the proportion of sulphur, salt, and the mercury they contain. Copper is highly colored, and it was said to contain much sulphur while iron was supposed to contain an excess of salt. The sulphur-mercury-salt theory was then championed by Paracelsus, and the doctrine gained acceptance among the alchemists. Salt, on the other hand, has been considered to be a less important principle than either mercury or sulphur.

The same idea found to be underlying these doctrines came much later in Stahl's phlogistic theory during the eighteenth century. It attempted to account for the combustibility of bodies by assumption that the bodies contain "phlogiston" which was the hypothetical principle of combustion. The concept of "phlogiston" approached the modern idea of an element than the alchemistic elements or principles. It was more evident later in the history of chemistry that the properties of chemical substances were not conferred of certain elements entering into constitution.

It was interesting to note that the alchemists combined the theories with Aristotle's theory of the elements which was earth, air, fire, and water being more interior than the principles whose source was to be the same elements.

According to Sendivogius in Part II "The New Chemical Light" that the three principles produced out of the four elements were as follows: nature whose power is in the Will of God was ordained from the beginning that four elements should act on one another so fire began to act on air and also produced sulphur, air acted on water and also produced mercury, and water by the action of earth produced salt. Earth did not have anything to act on and did not produce anything, but instead became the womb of these three principles. The three principles for which the ancients mention only two was clear that they omitted the third which was salt.

Within these coverings of outward properties was hidden the secret of all material things. One alchemist writes that the elements and compounds were made of a subtle substance or humidity by diffusing the elemental parts and preserving things in vigor are called “The Spirit of the World” which proceeds from “The Soul of the World”. It was the one life that gathers all things so the intellectual, celestial, and corruptible formed the one machine of the whole world. It was not necessary to point out how this approached modern views regarding ether in space.

The alchemists felt the metals growing from the womb of the earth and the knowledge of the growth being important. Thomas Norton has his own opinion among the alchemists. He felt each and every thing has its own environment that is natural to it. They had the belief in the growth of metals and that mines were closed for awhile in order for the supply of metals could be renewed.

The fertility of Mother Earth forms a subject of an illustration in "The Twelve Keys" of Basil Valentine. He felt that the power of earth produces things that grow forth from it, and those who feel that earth has no life makes a statement that is contradicted by ordinary facts. If something is dead, then it cannot produce life and growth which leads to the quickening of the spirit. The spirit is considered to be the life and soul that dwell on earth, and it was nourished by heavenly and sidereal influences. For herbs, trees, and roots as well as metals and minerals receive growth and nutriment from the spirit of the earth which was also the spirit of life. The spirit was fed by stars and was capable of imparting nutriment to things that grow. The minerals were thought of as hidden in the womb of earth and nourished with the spirit she receives from above. He felt that the power of growth was not imparted

by earth but by the life-giving spirit. If earth did not have this spirit, it would be considered to be dead and not able to afford nourishment. The sulphur would lack the quickening spirit which there cannot be neither life nor growth.

The idea that there was a growth of each metal was under the influence of heavenly bodies. It was a theory in harmony with the alchemistic view how unified was the Cosmos. The metals were referred to by the names or astrological symbols of peculiar planets. The metal gold was associated with the Sun, silver was associated with the Moon, mercury was associated with the planet mercury, copper was associated with Venus, iron was associated with Mars, tin was associated with Jupiter, and lead was associated with Saturn. It was thought that due to observances of astrological conditions that it was necessary for carrying out alchemistic experiments.

The alchemists regarded gold as a perfect metal where silver being more perfect from the rest. Gold was considered to be the most beautiful of all metals. It retains its beauty without it being tarnished. It has the ability to resist the action of fire and corrosive liquids, and it is not affected by sulphur. It was regarded as the symbolical of the regenerate man. Silver is also beautiful which wears well in pure atmosphere and also resists the action of fire, but it can be attacked by certain corrosives such as aqua fortis or nitric acid and also by sulphur. From all metals from one seed in nature works continuously up to gold. They felt that all metals were gold in the making. It was their existence that marked the staying of nature's powers. Mr. Philalethes states that the metallic seed is in the seed of gold. Gold was thought as an intention

of nature with regard to all metals. They felt that if the base metals were not gold, then there must be an accidental hindrance and they are all potential gold.

Another alchemist puts it this way. The substance of metals was one in the course of time; the foreign and evil sulphur of baser metals by gradual digestion changed by its indwelling sulphur into gold. This was the goal of all metals and the intention of nature in the mineral kingdom, vegetable, and animal kingdoms. Nature demands a gradual attainment of perfection and a gradual approximation to a high standard of purity and excellence. This was the alchemists view of the generation of metals. This was the theory of utmost importance which gave the idea that all varying forms of matter have been evolved from some one primordial stuff. This was a chemical principle for which chemical science lost sight of. Alchemists were great evolutionists, but they did not make a mistake regarding the evolution as explaining away the existence of God. The alchemist recognized the hand of the Divine in nature. Of course, in modern science we can't accept this theory of the growth of metals, but we can appreciate and accept the fundamental ideas behind it.

The alchemist worked hard to assist nature in gold-making. The pseudo-Geber taught that imperfect metals had to be perfected or cured by applying medicine. There were three forms of medicines that were distinguished. The first was to bring about a temporary change and changes thought by the second class having not been complete. He writes the medicine of the third order and calls every preparation when it comes to bodies with its projection and take away all corruption perfects them with the "Difference of all Complement". This would be considered to be the true medicine that produced a real and permanent transmutation to be The Philosopher's Stone which was considered to be the masterpiece of alchemistic art. There were similar views by other alchemists, but still others taught that it was necessary to reduce the metals to their first substance. There were often two forms of The Philosopher's Stone which have been distinguished or two degrees of perfection in the one Stone. It was considered to be the transmuting of the "imperfect" metals into silver being white, the stone or "powder of projection" for gold to be the red color. The medicine was described as a pale brimstone hue.

Most alchemists who claimed knowledge of The Philosopher's Stone or the material prima for its preparation kept its nature secret and spoke only in the most enigmatical and allegorical language. Most of the recipe meanings contain words that have unknown meanings. Gold and silver was employed in making the medicine, and after projection was made again in metallic form the alchemist imagined the transmutation has been effected. There were just a few recipes that were intelligible which the most that could be obtained by following out the instructions to be white or yellow metallic alloy superficially resembling with silver or gold.

The mystical which is distinguished from the pseudo-practical descriptions of the stone and its preparation were far more of the interesting of the two. According to Paracelsus work on The Tincture of the Philosophers, all we have to do is mix and coagulate the "rose-colored blood from the Lion" and "the gluten from the Eagle" by which he meant combine "philosophical sulphur" with "philosophical mercury". It was this opinion of The Philosopher's Stone that consists of the "philosophical sulphur and mercury" combined to perfect unity and they liked this union based on the conjunction of the sexes of the marriages. Eirenaeus Philalethes explains that the preparation of The Stone was necessary to extract the seed of gold, but this could not be subjected to gold with corrosive liquids, but only by a homogeneous water or liquid which was known as "The Mercury of Sages".

According to "The Book of Revelation" of Hermes which was interpreted by Theophrastus Paracelsus concerned the Supreme Secret of the World, the Medicine, which was identified with the alchemistic essence of things or Soul of the World. In other words, it was the spirit of the truth which the world can't comprehend without the holy ghost or without those instructions who knew about it. The same is a mysterious nature with strength and boundless power. The spirit was named "The Soul of the World". The soul moves limbs of the body and there was soul in the limbs of the body. The spirit is in all elementary created things. Many have sought it, but only a few have found it. It exists in everything and every place at all times. It has powers of all creatures and the actions found in all elements, and qualities of all things even for the highest perfection. It also heals all dead and living bodies even without medicine and has the power to convert all metallic bodies into gold.

Mystics have been overfond of ascetic ideas such as the development of the soul could only be possible with the mortification of the body. All true mysticism teaches that if we could reach the highest goal possible for man along with the union with the divine, then there must be a giving up of our own wills which is the abasement of the soul before the spirit. The alchemists taught for the achievement of the magnum opus on the physical plane. They needed to strip the metals of outward properties to develop the essence within. Helvetius states that the metals were hidden in their outward bodies as kernel was hidden in the nut. There was every earthly body whether it is animal, vegetable, or mineral kingdom. The earthly bodies was the habit and terrestrial abode of the celestial spirit or influence which was the principle of life or growth. The main secret of alchemy was the destruction of the body which helps the artist get at the living soul. The killing of the outward nature of material things was brought about by the processes of putrefaction and decay. This was the reason why processes figure largely in alchemistic recipes for preparing the "Divine Magistery".

The alchemists used the terms "putrefaction" and "decay" indiscriminately which applied them to chemical processes. There were pictorial symbols of death and decay representative of such processes to be found in alchemistic books. According to "A Form and Method of Perfecting Base Metals by Janus Lacinus, the Calabrian which is a short tract prefixed to "The New Pearl of Great Price" by Peter Bonus shows interesting images of plates. A picture shows a palace of a king (gold) where we observe him sitting crowned upon his throne and where he is surrounded by his son (mercury) and five servants (silver, copper, tin, iron, and lead). Another picture shows the son incited by his servants kills his father, and then in another picture he catches the blood of his murdered parent in his robes where we understand that an amalgam of gold and mercury was to be prepared. The gold disappears and dyes while the mercury was colored. Another series of pictures shows a grave being dug. A furnace is made to be ready. Another picture in the series shows the son to throw his father into the grave and then leave him there but both of them fell together. The sixth picture shows the son being prevented from escaping and both son and father being left in the grave to decay. The instructions in symbolical form is to place the amalgam in a sealed vessel in the furnace and allow it to remain there until some change is observed. The father is restored to life which shows the symbol of resurrection of frequent occurrence in alchemistic literature. By carrying out this resurrection, we see that the gold will be obtained in a pure form. It is now the "great medicine" and in the last picture the king's son and his five servants were made kings based on their powers.

The alchemists felt that minute proportion of the Stone projected lots of quantities of heated mercury, molted lead, or other base metal that would transmute the whole into silver or gold. The most minute quantity of the Stone was sufficient to transmute lots of quantities of "base" metals has been ridiculed. Some claims of the alchemists are out of reason. The disproportion between the quantities of Stone and transmuted metal can't be advanced as an "a priori" to the alchemists claims just as much as a class of chemical reactions. These chemical reactions were known as "catalytic" where a small quantity got some form or forms exist. As an example, cane-sugar in aqueous solution can

be converted into two other sugars by small quantities of acid. As for sulphur dioxide and oxygen, they will not combine under ordinary conditions. But they do so in the presence of a small quantity of platinized asbestos which was obtained in unaltered conditions after the reaction was completed and can be used over and over again. The process is also employed making sulphuric acid or oil of vitriol. The catalytic transmutation of chemical elements was also considered to be a conjecture.

The Elixir of Life was described as a solution of the Stone in spirits of wine or identified with the Stone itself could be thought of as under certain conditions to the alchemist that it would restore him to the flower of youth. The idea, even though not infrequently attributed to alchemists, that the Elixir would endow one with a life of endless duration on a material plane. It was still not in strict agreement with alchemistic analogy. The effect of the Elixir was physiological perfection which ensured long life, and it was not equivalent to endless life on a material plane.

According to Paracelsus, The Philosopher's Stone purges the whole body of man and then cleanses it from impurities by introducing new and youthful forces which it joins to the nature of man. Another alchemist states there was nothing that might deliver the mortal body from death. There was still one thing that postpones decay, renew youth, and also prolong the life of a human being. The theory of a solution of The Philosopher's Stone was also thought to be a species with gold. It constituted The Elixir Vitae that can be traced to the idea that gold in a potable

form was a cure-all. During the latter days of alchemy, any yellow-colored liquid was foisted upon a public as a medicinal preparation of gold.

Let us step back and look at some practice methods used by the alchemists. The alchemists worked with large quantities of material compared to what we use in modern times. They felt that time was important and gave a desired change in their substances, and they were to repeat the same operation such as distillation on the same material over and over again. This demonstrated their unwariness patience even if it effected little attainment on their end. They paid much attention to any changes of color they observed based on experiments and the descriptions of supposed methods to achieve the magnum opus contained detailed directions to various changes of color which was obtained in the material operated if there was success to the experiment.

According to Espagnet in his "Hermetic Arcanum", he felt the means or demonstrative signs were colors that were successive and orderly affecting matter and the affections and demonstrative passions that it follows. There are three special ones that are critical to be noted. Some even add a fourth. The first is black which was the Crow's head because its blackness whose crepuscular sheweth; the beginning of fire of nature and solution, and the blackest midnight sheweth the perfection of liquefaction and confusion of elements. The grain putrefies and it was then corrupted and was more apt for generation. The white color was given the perfection of first degree and of white sulphur. This was

considered to be the blessed stone. The Earth is white and foliated where philosophers sow their gold. The third is orange color which was produced in the passage of white to red as being in the middle and being mixed as both as the dawn with his saffron hair which was the forerunner of the sun. The fourth color is ruddy and sanguine which is extracted from white fire only. Whiteness is easily altered by any other color, but the deep redness of the sun perfected the work of sulfur which was called sperm of the male, the fire of the stone, the King's Crown, and the Son of Sol.

Besides the decretory signs that were found in matter and shew its essential mutations, almost infinite colors appears and shew themselves in vapors as the rainbow of the clouds. These quickly pass away and were then expelled by those that succeed which affects more of the air than the earth. The operator needs to have gentle care of them because they were not permanent. They proceed from the fire painting and fashioning everything after pleasure or casually by heat in slight moisture. This was not without a mystical meaning as well as a supposed application for the preparation of the physical stone.

Specific apparatus was used to carry out experiments during this time period. A furnace and alembics were used. The alembic proper was still-head which can be luted on a flask or other type of vessel. It was used for distillations. The alembics were employed in conjunction with the apparatus for subliming different volatile substances. Another apparatus used for sublimation consisted of a type of oven and three detachable

upper chambers called aludels. In both forms of apparatus, the vapors were cooled in the upper part of the vessel. The substance is deposited in the solid form which is purified from less volatile impurities. Another apparatus was the anthanor or digesting furnace and a couple of digesting vessels. The vessel was employed for heating bodies in a closed space where the top being sealed up when the substances to be operated had to be put aside and the vessel itself heated in ashes in an athanor where a uniform temperature was maintained. The pelican was used for a similar purpose where the two arms were added so the vapors would be circulated.

Chapter 7

MAGIC OF ALCHEMY

Some people felt alchemy to be a magical secret and now lost to others. Others say it was a fraud of charlatans. Even others felt alchemy was some occult philosophy in which metals and their transmutations were elaborate symbols. Even others had different views such as being a forerunner of modern experimental chemistry. In fact, all of these views were correct. There were many kinds of alchemists as opinions of the art. There were some who were adept of great wisdom and the evidence was strong that some of them changed base metals to gold by mysterious tinctures. There were still those fools who spent fortunes in their lifetime on theories and experiments. Yet there were still philosophers whose doctrines who were just as pure as the symbolic operations for which they were described. There were still many frauds out there as well.

During its time period, alchemy held the layers to great secrets of matter and spirit and it still has a message for modern life. Alchemy is all about gold that was ruled for hopes, fears, and ambitions of countless generations. Alchemy, in terms of the scientific perspective, was concerned about transmuting base metals into gold. Alchemy, in terms of philosophy, was concerned with the perfection of

human spirit which was affirmed by the scientific act of transmutation. Some of the experts succeeded in both spiritual and material operations. Gold has been considered the perfect and precious of metals. We remember many sacred and beautiful objects made from gold and even with the passing of thousands of years, gold's beauty and value has never declined. Even "The Golden Mask of Agamemnon" discovered during The Bronze Age site of Mycenae in Southern Greece was created between 1550 - 1500 BC. It was found in a royal cemetery, and it has been described as the "Mona Lisa of Prehistory".

Gold was thought of being natural as a symbol of the perfected human soul. Gold can be purified in a series of stages. It was even thought that the ancient Egyptians buried their kings with golden ornaments so the soul would have powerful support in its progress through the halls of the dead. The wise men knew that gold would endure the great journey from life through death was the perfect soul. There were even robbers who defiled the tombs, stole golden ornaments, and used dead kings as firewood. A later prophet advised men to lay treasures in heaven. During the time of King David, the Psalms have spoken of the judgements of the Lord as time and righteous such as the "more to be desired than gold, you than much fine gold." There was also a time when gold was a favorite symbol with the gurus of ancient India who taught that there is one great soul in all of us. They felt that gold could be moulded and hammered into a variety of shapes and different ornaments and still remain gold in all its manifold forms. Alchemy came about as the spiritualization of ancient craft and chemistry of refinery gold.

The philosophy derives from Eastern religions and taught that the solid of the material world was the same, only the soul survived death, and the outward material could be transformed and then refined. Baser metals would be transmuted to the purer form of gold and this act of transmutations grew from the mystery of the relationship of material and spiritual life.

Alchemists wrote about the things and used secretive language concealing as much as they remembered. Only the experts of pure heart and understanding could follow mysterious hints and allusions hoping to learn the mystery of making gold on both material and spiritual planes. The adepts called it "The Great Work". Eighteenth century naturalist Buffon wrote his article on gold in Histoire des Mineraux

"It must be admitted that nothing can be got from books on alchemy.

> ***Neither the Table hermetique, The Table des Philosophus, nor Philaslethuse, and various others whom I have taken the trouble to read and study seriously, have offered me anything but obscurities and unintelligible processes."***

The book explains the material and mystical aspects of alchemy and its transition from alchemy to modern chemistry. Older alchemists, who transmuted both matter and spirit, was a grand synthesis we must rediscover. There is truth in the claims of the fourteenth century from master alchemist Nic Nicholas Flamel who wrote:

> ***"And then afterward I did so will, the Red Powder upon a like quantity of mercury, in presence once more of Perrenelle alone, in the same house on the twenty-***

> ***fifth day of April about five o'clock in the afternoon, which mercury I verily transmitted into almost as much pure gold, very certainly better than common gold, suffer and more flexible. I can say this with truth. I have this perfected it with the help of Perrenelle, who understand it as well as I myself."***

Nicholas Valois followed him and claimed equal success and added:

> ***"Which thou wilt do as I did it, if then wilt take pains to be what then shouldest be- that is to say, pious, gentle, benign, charitable, and fearing God."***

Gold remains the dominant influence and reference point of whole material life. There was some mysterious occult significance to all of this. It is a time to recover the lost knowledge of the metaphysical equivalent of gold that gave meaning to life beyond the passions of wealth and ownership. Our understanding of the perfection of material science and insights and former concepts of former philosophers may be refined into the purer metal, and such new Alchemy can be considered to be Great Work.

But remember alchemy was considered to be a journey to the unknown. Within the hidden chambers of the realm of alchemists, there lied a profound journey which transcended even the great boundaries of the chemistry laboratory. It was considered a quest which stretches far beyond from the unknown to the expanding consciousness. To the alchemists their art was not an exploration of matter of elements but also the goal of transmutation of this enigmatic pursuit into the ethereal realm of God-Consciousness. They felt there was a higher plane of ancient wisdom and

divine knowledge that converged. The alchemists worked hard to unlock the secrets which lie beyond ordinary perception by using their conscious minds to higher states of awareness and also to explore higher dimensions. They take the time to tap into the transformative power of changing perform forms within archetypal worlds. During the ethereal odyssey, the transmutation process was not confined to the boundaries of the physical world. They traverse transitional states from the consciousness and encounter a plethora of psychic thoughts which lie beyond the realm of ordinary senses to extraordinary levels revealing extrasensory perception.

Soon they become the masters of energy manipulation carrying out the unseen forces to shape the reality of the world. The mind becomes a strong force and becomes the catalyst for alchemical transformation. The alchemists themselves tap into the realm of subatomic particles, command, and bend their properties at their will. Between this consciousness and matter, intentions become manifestations of their desires at the core of alchemy which lies the concept of prima materia. It is this primal and essential substance which underlies creation. The divine clay where the universe contains infinite potential and inherent possibilities for each individual. As alchemists delve in the depths of prima material, they seek to unravel the mysteries and discover the keys to transformation and the secrets of the cosmos aligned with rhythmic pulse of the cosmos.

The alchemists felt that the pursuit of perfection can accelerate the natural process propelling themselves to higher states of being. In the sacred union of

the cosmos, they too become the co-creators harmonizing the threads of existence and weave new patterns of transcendence. It was the alchemist's true laboratory which lies within the sacred sanctum of the soul. It is this place in the depths of their being where a profound transformation takes place. It was the alchemist that embarks for voyage to selfdiscovery, growth, and transmutation. The crucible of the spirit becomes the vessel which holds fire of transformation that refines the essence and transmutes the leaden aspects of themselves into radiant gold.

The mysterious realm of alchemy takes place where science and spirituality intertwine. The alchemists unravel the enigmas which has perplexed humanity for ages through dedication and understanding and harnessing the principles of transmutation. They embark on a path of spiritual and personal metamorphosis. They felt they hold the path and promise of unlocking the secrets of existence as they embark on the esoteric sciences of alchemy. Keep the following key points in mind about alchemy. These key points are The Philosopher's Stone, The First Matter, The Seed, Mercury-Sulphur-Salt, The Seven Metals, A Universal Panacea, and The Fire itself.

We will be discussing more about The Philosopher's Stone which lies within the labyrinthine depths of alchemy. This transformative catalyst transcends the realms of matter to elevate objects to ultimate expressions. The power is unparalleled and capable of transmuting base metals into precious gold and human beings into divine beings. They seek the key that unlocks states of regeneration and resurrection. It has been said that legends and ancient texts from numerous cultures bear witness to mystical stones. Taoists enlighten seeker of wisdom and honor the stone as “The Stone of Wisdom”. The Burmese associate the stone with the ethereal “Stone of Live Metal”. In the Islamic tradition, it was found that veneration is bestowed from a meteorite fragment within the Kaaba which has turned black due to humanity’s sins. The mythical city of Shamballa guards the precious jewel called the “Cintamani”. The love jewel symbolizes the perfect expression of the heart chakra which also enlightens the mind. The stones have profound meaning that transcends the physical to spiritual principles within the depths of existence. The alchemists regarded the philosopher’s stone not as a literal stone. It considered not to be a tangible object which someone holds one hand, but it was instead a mystical concept composed of the first matter. It is the harmonious blend of four elements hidden within the depths of humanity. The stone possesses divine powers which transcend the physical realm, but the appearance can be deceiving. It was commonly found as filth, unattractive, and disregarded by the uninitiated. This is what we call the paradox of the stone. Nature concealed beneath the veil of its outward appearance.

Love was considered to be that potent force which radiates from the core of human expression and acts as a transformative catalyst from the mundane to the divine. Love had the power to purify, illuminate, and also transmute the body, mind, and soul intertwined with the essence of the stone. Even beyond love, the divine gives wisdom, power, and light intertwined with the essence of the stone. These qualities possess an extraordinary ability to purify and illuminate deep recess of an individual as seekers continue to embark on the magnum opus. It was their journey of self-discovery which allowed them to experience and comprehend fixed elements of existence. The fabric of reality and the absolute saying "Man, know thyself" becomes the guiding mantra that aligns with the divine spark, their divine self, and christ emerges as the embodiment of the philosopher's stone. The cornerstone holds the power to transmit human existence to bring fourth spiritual transformation and enlightenment within a human being. The heart and crown chakra serves as focal points where the stone unleashes its transformative power anchored within the heart. It is the divine flame of life that radiates energy and magnetism which is nourished by the mystical stone and the love, life, and light principle that extends throughout the inner workings of the human body.

In return, this bestows the individual sense of youthfulness of vitality and health. Yogic traditions of ancient Hindu sages echo wisdom and proclaim that those devoid of love should age rapidly. The intricate mechanisms of the stone's power reveals themselves a portrait of this transformative potential.

Occultists propose the philosopher's stone embodies a gnosis. It is this transmission intertwined with mystical and occult principles that facilitates perfection within a single lifetime. Initiation and ritual play important roles during the days of alchemy. The presence of an adept transmutes those fortunate enough to encounter them as embodiments of the philosopher stone. We will encounter the wonders of the philosopher's stone in the next few chapters. We will find it as an enigmatic symbol of transformation and transcendence in the strange

realm of alchemy. It reaches across cultures and beliefs which embody the power to transmute the mundane into the sublime and elevate humans to the realms of the divine. It is the eternal quest for self realization and the workings of hard truth. The journey towards the philosopher's stone gives promise metamorphosis and a glimpse into the boundless potential for the human spirit. But before we can discuss in more detail about the philosopher stone, we need to go back in time and learn about the origins of chemistry before alchemy came about. Little records have survived during ancient times during the beginning of civilization. As you continue to read the chapters, we will build up to the point and talk more detail about alchemy, mystical sciences, the philosopher's stone, and the final alchemists which led to the beginnings of modern chemistry. But we are going to start with first the precious substance of all which is water.

Chapter 8

THE ALCHEMY OF WATER

There are many known substances out there, but the most unusual substance and by far the most cryptic element has to be water. We will explore together different alchemical meanings behind water and the connections of water to both ether and mercury. The fluid motion water takes on has meaning to it and has an expression of water's high nature in the environment. Water was known as the mystical and esoteric form known as the prima materia. This was the prime substance that was transmuted. It was the prima materia that was the prime material the alchemist used from the beginning of some of the great work along with the processes of transmutation to turn the end result of the philosopher's stone or gold. The prima materia was also known from the beginning and the end of what is being transmuted for the entire process. It has been called so many different names, and this prime material has been shown for every single type of symbolism. The most telling name was fire water which was also known as mercury. It was also known as quicksilver. It was also known as the alpha and omega symbolizing the beginning and the end product such as the living water spirit or chaos dragon fire.

The prima materia was important to alchemy and to the metaphysical terrain of the entire universe. Knowing the prima materia was essentially the higher nature's quality of water. Water's fluidity expresses about the nature of water but also the nature of water that corresponds to which was the prime substance. The fluid nature of water and how it traverses the many different states of density from the solid state to a liquid state and then to a gaseous state. Water encompasses different phases of matter which depends on the phase water will be occupying and will determine what water will be referred to in that phase.

The alchemical property of water was considered to be the prime material but other properties of water encompasses whether it exists as the solid state, liquid state, or gaseous or vaporized state. In the vapor state, water was known as the essence which comprises the soul, and in the liquid form water would be in the prime state because the fluid motion would be its true nature. It also corresponded to the metaphysical properties of water which serves as the subconscious. This explains why water has been correlated with emotions and its subconscious processes. Water in this form serves as an intermediary between the unconscious or the soul and the conscious mind of the solid particle state. Water has been known in alchemy to serve as the spirit but it has a fluid motion showing its nature which moves and communicates important messages and information from the celestial or astral realm to the physical and terrain realm. It goes through the unconscious to the conscious mind where water in the natural form was thought of as a fluid form that goes between worlds. Water embarks and delivers many important messages

between the different realms which could not otherwise correspond and also communicate with one another. This explains why we can call water in the liquid state as quicksilver or even better known as mercury as an important form of planet to communicate. It is the process of communication which is what water does best even in its most inherent and intrinsic forms. It was not an accident that we have certain archetypes emerged from the fluid motion of the subconscious state in a specific form. Hermes was known as the father of alchemy, and he was also known as a psychopomp. He relays information from one realm into another realm, and this is exactly what the subconscious does, and this is done through fluid motion to traverse all realms including the negative and positive attributes.

Water was known to be malleable, and it was shown through the physical motion which it takes on as a fluid. Remember we are taking water to be the highest of all chemical forms as the prima materia which is what all of the cosmos came from. It was also known as the prime substance.

It was the primordial activity or substance itself that divides and continues to divide and again in a creative process known as the prima materia. It was known as both its spiritual and elemental form as mercury and also as Mercury's Earthly form to be water. All three of these states belong to what we call the same trinity. This explains why during the days of alchemy that Mercury was also known as the mother of the stone since it was the medium of worlds.

Water would be the messenger of the Gods for moist and dryness depending

on its state. Based on life and death which would be the breath of the cosmos, mercury symbolizes itself as both the spirit and the mind.

It was considered its counterpart as a metaphysical expression of water. It was the division of what happens of the prime substance considered as the original oneness. From the original oneness or the primordial substance, it carries out a wave motion but not in its fluid physical form as it is observed on Earth. It was thought of as taking its motion as a wave motion of the electromagnetic fields. The energy the material form is made up of and created is what makes water to be the prime material. This is what the cosmos is made up of and below it divides again as water being the wave motion we a quantum field and a waveform state. The solid form becomes collapsed to its metaphysical form and then turns to the waveform. Water was considered to be the waveform state, and ether was water's higher corresponding element. Ether was known as being one undetectable and would be considered to be The Fifth Element and known to be the quintessence. The Fifth Element makes up of other elements within itself, while water divides itself again for creation. We have the higher expression of water which was known as ether and ether was made of all the elements and one of the elements ether was made of water was water. This shows how water divides itself further into components because water was thought to be found in a higher nature.

There was also a middle nature and a lower nature that had corresponding features of every level of reality. We know now that water on Earth has its form which corresponds to a higher nature. Again, it has a tendency to divide itself into the electromagnetic fields which consists of electricity and magnetism just like the masculine and the feminine energies. Ether divides itself into the electromagnetic field and a waveform state, and then it continues to divide until you have the realm we live in now based on the physical properties. They all come from a higher nature which was ether in many different forms, but its primary connection has to be with the element known as water. By the time water takes on the terrestrial form, we know water as a liquid we use in our everyday life. It does not represent the metaphysical components that was cryptic and everything that has creation within it. This explains why water was known as the universal solvent because water can dissolve almost anything and what water cannot dissolve explains what it cannot do. Our skin forms barriers just like the buildings have its structure not being able to dissolve water. Nevertheless, water was known as the universal solvent.

Remember that water has the ability to dissolve which is due to its law of "like attracts like" and gets neutralized in a chemical process. In reality, there are positive and negative charged chemical compounds in water that goes around attracting and repelling other things in a neutral state. This explains why this neutral state was known in religion and in spirituality as being a purifying force.

It is water's ability to dissolve impurities through a neutralization process and creates what we know in our psyche or in the spiritual community as purification. This also explains why water is used to baptize. This also explains why water can be imprinted on by using the vibration of our voices and by doing so it shapes water or by using our own mental impressions to send water. Water's fluidity allows us to take the shape of what we intend to use it for. During the days of alchemy, water was thought to contain all information, all stardust, all elements, and all molecules one could think of within the structure itself. When we impress upon it, we are recording our intention onto it. It was thought of as a good recorder, its ability to store memory, and why it is knowing the new age and the spirit of this age as storing memory within it. Water has all the information of the cosmos inside it. We activate water by calling forth all the information that is contained within it. We are imprinting our intentions on water, but what we are doing is summoning the activation of all the parts within it so we can impress upon the water and so it can take on its fluid motion.

This is how we can see water that functions as the emotional activity that is stored within the subconscious realm, but it is also the activity in the unconscious realm so it does act as the unconscious and the subconscious. It is just in different states. This is why we say water is in its quicksilver form or mercury when it is in the subconscious realm, and when water operates in the unconscious realm it is what we call the prima materia.

When water corresponds to the ether or what we call the unconscious state the ether contains all of the elements. The unconscious state contains all the elements based on how ether can be expressed as plasma being a gas and how this ties in the properties of water. Water also takes on its gaseous state, and when it is in the gaseous state it comprises the soul so when it is known in the vapor state we can look at it as an essence and the essence in physics we look as a gas.

When water is in the gas state or vaporized state, it is the ether that is expressed as plasma since plasma is a gas. We then go a layer in and water is expressed as the subconscious which was also known as the electromagnetic field or the waveform state. In this state, water can be expressed as the subconscious but also in terms of emotions. Emotions are thought of as energy in motion and water in its solid state was also known as the alpha and omega state.

Water contains all of it, and it expresses it in any form given depending on the realm water operates within. Water takes on the role of different worlds that it occupies but also water has different meanings because of how important it is and how fluid its nature is. When we hear about the living waters or drinking about waters instead of poisonous waters, it does not refer to the literal sense of the term water. It was speaking of one of water's important natures which contained wisdom so water was also known as being referred to as knowledge and also being referred to as wisdom when we hear about living waters.

Water also takes on an extra role being referred to as true knowledge and true wisdom that leads into consciousness and then into ascended states of consciousness in comparison to other waters which mean other knowledge that leads a person away from consciousness and away from true gnosis.

As stated earlier, water divides itself into polarities to create further down into the physical realm. This is what we mean when waters were separated from waters or waters of this realm

were different from the waters of other realms. If water was the primordial substance, the prime material being used to create what was happening were waters of a higher nature and waters being separated into a lower nature. This is what we refer to as astrolite so you have the astral realm being divided into the lower astral realm and then into the higher astral realm. You also have the terrestrial world being divided into the Heavenly Realms and also in to the Earthly Realms. You also have this divided from the firmament and the firmament. There is this constant division of waters that is being used to create the universe.

Water has its true nature as the emptiness that was being vibrated so fast that it was perceived as not vibrating at all as the chaos which contains both spirit and matter is both darkness and light. Water is where the astral realms were considered to be called and the consciousness was also known as the Deep Waters where it pulls from its own water to create the astral waters. It was this medium that ancient civilization was carried out on. We know that currents apply to the ocean or even electrical currents. They are the intelligence of both the energy and motion of consciousness. Currents were thought of as fundamental to the nature of reality itself. They were the activity of consciousness which manifests as light, and in fact, light was regarded as the electrical emissions or currents. All of this emanates from the eternal waters of the prima materia.

A strange and esoteric meaning of water was thought of as light since you have light that divides itself into light for creation. Light travels on the ether in wave forms so light travels so light travels as a waveform and the there is the ether that travels as a backdrop. Light travels upon or gives the illusion of traveling upon the ether and both consist of waveforms.

The lower nature is water, and the higher nature is the waveform state what is creating on. It creates the wholeness right and pulls from its wholeness to create a process of pulling the waveform state, but it was also known as vibrations so the waveforms or the light is traveling in one direction an then counterbalancing an area that moved from opposite and equal ways. In order to create a waveform, there has to be wholeness and to create anything from wholeness we need movement and vibration which explains why vibration is created the way it is. As an example, sound creates our reality and when you have a vibration, then there becomes a movement away from wholeness and then there is a counter balancing movement away from wholeness and then another counter balancing movement and the pattern continues, and this vibration is the waveform state. Vibration is created which is a counter balancing act which consciousness takes on in order for consciousness to create and all of this has to do with cryptic and mysterious substance what we call water.

Speaking about water, the water chakra has been known to be connected to other energy centers in the body especially the root and sacral chakras. When the chakras are balanced and open, they then work together to promote emotional stability, creativity, and then a sense of flow for life. There is more to talk about mystical chakras coming up next.

Chapter 9

MYSTICAL CHAKRAS

The alchemists considered the chakras to be the "seals of the planets" or the "furnaces of the soul". The speciality of alchemical teachings was that there were three levels of energy for which chakras took effect. They were considered to be the nigredo, albedo, and rubedo. The energy centers found in our bodies are best known by what we call the chakra system. It has to be one of the most core and also a foundational part of spiritual knowledge. We are still continuing our understanding of alchemy but in terms of the chakras. We are going to look at what each energy center is all about. The term chakra comes from Sanskrit. It has ancient roots that go back into satanadharma. Chakra can also mean wheel and has been known as Circle or cycle. It was also known as a portal, and it was known as a vortex so there are different ways looking at this energy center concept. We can look at it like a vortex and a portal as well as a cycle so there was some revelation in a wheel that happens inside the energy centers and from different walks of life. We can view the chakra system within our body so this does not have to be an eastern concept. We can objectively perceive this.

There are more than seven energy centers that take place within our body. Some of the ancient people perceive it to be the chakra network as 144. Others perceive it as 14,and yet others perceive it to be 13 or even 12.

There are energy centers that got every way throughout our energy body but also throughout our whole reality. The main energy centers run energy through and project ourselves through inside space and time and reality itself. The classical seven chakra system starts from the base of our spine with the root chakra.

This is known to be linked with our security, our ego, and our identity as an expression of our survivalism so our ego consciousness in the most base form form means that there is truth and reality based on our identity as an individual to the survival and security here on planet Earth. It contains wisdoms a lot of times. We view the center being less superior because it is connected to the Earthly real, but this is the core when it comes to our ability to thrive on Earth. We find that the root center becomes connected to our health, and it is connected to our ability to also heal and connect to our ability to receive abundance. In order to receive abundance have have abundant consciousness, we need to have security and safety so we can thrive and have the ability to attract abundance or allow and welcome in abundance. Even including those in ancient times, we need to have a healthy root center and since we know that abundance can be connected with abundance consciousness, it can also be connected with the opposite of abundant consciousness which is scarcity consciousness or what we call poverty consciousness. It is like describing finances and money

because that is directly tied our ability to survive and have security over here. The root center carries core themes that deal with us being in this reality but also have to do with the most essential and core by itself.

Wisdom from this root chakra has to do with wisdom of our ability to survive in the world and from our individuated self-expression as well as feel safe and secure enough which opens up whole energy fields having the capacity to receive the ability to from survival to again survival.

The second chakra is considered to be sacral chakra, and this sacral chakra is located below the naval region. Below the Naval region, we have our creative identity so all of the different energy centers that are within our body have their own wisdom and they also have their own identity. The sacral chakra the ancient people would look at as its core identity because it is the creative energy center and as a creative energy center linked with our sexuality.

The sacral chakra is symbolized through the water element because it is considered to be a fluid. It is the creative and sexual identity because this creative energy is considered to be sexual identity. It is the emotional body and the emotional identity so this is considered to be a center that is linked with abundance, and it is linked with abundance in a more spiritual fashion rather than how the root chakra gets linked with the abundance. In the root chakra, abundance gets linked with the security in identity which is developed and the ego that is the identity's ability to survive and feel secure in the world. The sacral chakra is one that has an emotional and creative identity linked with the ability to manifest the security of the life those who live but then are

deeply linked with the ability to manifest through emotional body and through magnetism but also through the truth of how the ancient alchemists felt as creative and sexual beings. The energy center becomes immensely and extraordinarily creative. It is also linked with creative aspects of ourselves which embrace and bring those gifts forward into the world and manifests itself through a form of alignment.

The third energy center is known as the solar plexus, and the solar plexus region is located above the naval so in this center this is one that becomes directly linked with willpower. The center is about self-worthiness and selfesteem. The ability to feel confident so those can put themselves out there in the world. It is liked with ambition including the healthy and negative forms of ambitions. It is linked with vision, direction, the ability to have purpose and then go in the direction to move willpower in the direction of purpose so the center of the solar plexus region becomes important for the ability to put the truth of existence from the root center and from the sacral center.

It conveys the truth of lower centers and puts those out there in the world which corresponds to lower center so it is one thing there are root centers that want survival and to feel secure.

It does not know how to be able to harness moving in a direction in a world where they can achieve that comes into play with the solar plexus region. This solar plexus region becomes directly linked with the capacity to take acton and forward movement of what we want to manifest in the world. The solar plexus region becomes powerful when it comes to the ability to have permission to be who the ancient alchemists to be and to echo that to

the world. The solar plexus region has to do with visibility. It has to do with being seen so to have all the energy centers and then have all of our energy centers and then have that. The solar plexus region is thought of as a form of communication through action and through willpower. It is the truth of our entire being out into the world so it makes the energy center most responsible for making the alchemists visible.

The fourth energy center is considered to be the heart chakra. The heart chakra is located in the center of the chest. The heart chakra plays a role in the whole chakra system, and it is not the ability to be projected into reality, but also bridges the threshold between higher worlds of the throat center, the third eye center, and the crown center as well as between the lower worlds which is considered to be the solar plexus region, the sacral chakra, and the root chakra. The heart center is the bridge and threshold so it is a stargate between higher worlds and lower worlds which corresponds to the higher dimensions of being and the lower materialized dimensions of being. The heart center is the bridge, and it is known for being a bridge and it's known for being a bridge and the threshold it is known for being the compassion and empathy that lie since the truth of the heart center is oneness. We know that to give to another or to feel pain or to have care for the world. It comes from the knowledge and the gnosis that the alchemists were one and so the way the heart center is developed is through realization and through truth and the heart center is a bridge and the truth of being which is oneness, but it is the way that is developed to higher psychic capabilities. The great gifts that is activated and that exist within

is done through the heart center. If it is not done through the heart center, then it has to be done through the heart center, then it is done through bargaining, a force of manipulation, and that is not sustainable and won't last so the heart center develops the heart chakra is the psychic capabilities come back. It is through the first cultivating knowledge that the ability to know that the other is oneself and that oneness is considered to be the highest truth of the universe and having that to be the foundation from which the heart center is developed.

The fifth energy center is considered to be the throat center, and this throat center is the center of our throat. The throat center is vital for the entire chakra network because each energy center has their own identity and their own wisdom. It is the throat center that wisdom and all the identities within the energy field can be conveyed and are able to be out there broadcasted into the world through powers of vibrations. This is what the voice does when the throat center is used, and we are activating the vibratory field and so vibration conveys the truth, the voices how the truth is conveyed and how will another person be able to know the truth that is discovered from our root chakra.

How can the world know the truth of the sacral chakra and so forth is through the throat center so the throat center has a role within the whole network by being able to communicate to the outside world by what exactly goes on inside our inner world. The throat center is known as the truth because it does not matter how far the personal truths are in relationship to the world that surrounds us. The center is considered to be the portal of truth since truths are held within each one of the energy centers and their identities that communicates to the outside world. From there another person's throat center may eventually come back and forth and the truths will mix and mingle. One does not know what is communicated and conveyed through the throat center by coalescing the truths that is how the truth is purified. Even the alchemists learn new things, change their mind, or even stick stronger to values whatever that is the truth of their internal being meets the external world's truths, and furthermore the alchemist performs where the truth can be purified one way of another. The throat center is responsible for communicating the truth of oneself from all the different energy centers.

The sixth chakra and also the most popular chakra is Arjuna chakra which is the third eye center. The third eye center is located a bit above the eyebrows which is directly in the center of the forehead. This becomes the center that is linked without clairvoyance. This is the center that becomes linked with the ability to perceive the invisible or the metaphysical so the ability to perceive beyond the third dimensional physical realm. This has to do with other psychic abilities because the third eye region is where the other psychic abilities come. The heart center develops the emotional essence, the creative essence, and from that creative life force it develops in the heart center and then moves up into the third eye region and from the third eye region, this can be perceived as commingling about alchemical marriage or alchemical wedding that takes place. The third eye region is in the third ventricle in the brain. The ajna chakra which is the third eye is known for the ability to see the other worlds. It is the ability to

perceive the other worlds where the center is known for being the seat of intuition which is why the third eye has been known throughout ages being the seats of the soul. The seats of intuition is strongest within the energy center and from this energy center that intuition blossoms the strongest. The third eye center has been known for being where intuitive abilities and institution takes place.

The seventh chakra is known as the crown chakra which located above the head. This is the connection to the infinite. Infinity can be personified as the energy center. It is known to be a no thing, and this is where the eternal nature which is considered to be no thingness. The crown center is considered to be the silence of source. It is god-consciousness. It is the link with the Divine Essence so the energy center expresses the divinity the most. The crown center is the spiritual connection when those who talk a lot about how they connect with their higher guidance or connect with the higher self. They want to feel a sense and strength the crown center. Note that the crown center is not sensory so a lot of times those who want to feel their guidance and their higher self is what happens. They feel they want it to come through the crown center since that is the spiritual connection, but remember the crown center is not visible and becomes non-sensory so that puts a lot of work at ease.

The chakra stones are associated with specific color and energy which holds the inner balance and harmony. Positive energy flows throughout the body. The stones have been used during the days of alchemy and for centuries to promote healing, alignment, and spiritual growth. To summarize the root

chakra is the red jasper stone known for grounding and stability. It enhances feelings of security. The sacral chakra is the carnelian stone and is known for creativity and passion. It stimulates both inspiration and also emotional expression. The solar plexus chakra is the citrine stone which gives properties of confidence and personal power. It boosts self-esteem and assertiveness. The heart chakra is the rose quartz stone which is known for love and compassion. It fosters forgiveness and also empathy. The throat chakra is the aquamarine stone which is known for clear communication. It helps and facilitates honest expression and also effective communication. The third eye chakra is the amethyst stone which is known for intuition and spiritual growth. It enhances psychic abilities and also inner wisdom. Finally, the crown chakra is the clear quartz stone which is known for its connection with higher consciousness. It supports both spiritual enlightenment and also awareness. The effects of chakra stones is for balancing and aligning, absorbing and also transmuting negative energy, promoting emotional and physical healing, enhancing meditation and also spiritual practices, and supporting the universal well-being and harmony. Yet there was one important stone the alchemists really wanted to find which was known as The Philosopher's Stone.

Chapter 10

THE PHILOSOPHER'S STONE

It is now lost in hallowed antiquity, the goal of hermetic or occult science of alchemy has been recognized during at least the late Pagan world of the city of Alexandria which was the stone of the philosophers. The substance was known everywhere and nowhere. It as a stone that was not a stone to be capable miraculous magical feats. It might have had the ability to transmute base metals into noble ones such as gold. They also felt that could heal disease as well as transform mere stone to precious gems. Moreover, they felt they could reverse the process of aging. It could cause barren fields to sprout with the beauty of lush and life. They can imagine it to teach one to speak the languages of beasts and birds. There was even a version so subtle, then it could only be detected through taste and the version should allow one to observe and communicate with none short than the angels of the high heavens. It was remarkable that the philosopher's stone was the stuff of legends that are extraordinary but still enigmatic and obscure.

The exact origins of the concept and term philosopher's stone was equally mysterious. The concept was derivative

of two theoretical origins in Greek philosophy along with indigenous workings of Egyptian metallurgy. The first were the iliadic philosophers like Parmenides who held that regardless of all appearances the cosmos was a united ontological monad. It was one thing in topan. In other words, all is considered one. All that is different to the level of appearance was an expression of a deeper underlying metaphysical unity. The metaphysical concept was then put into a physical theory especially the generation of the elements and the metals by Aristotle in his text of meteorology. His book was popular in antiquity during the middle ages. Aristotle leaning on the impediclean theory of the four elements such as earth, air, fire, water, and the properties being dry, wet, hot, and cold. It is these elements that describe how the elements interact deep within the earth through vapors and exhalations based on the relationship to the sun that acts like a heating element as it rotates around earth. As the sun itself rotated around vapors or exhalations interact with simple bodies as he describes them in his book the physics and become fixed as elements or stoichaya or metals more specifically. These are fundamentally just ratios of four fundamental elements. They could be even decomposed and also converted to other elements which goes on the eternal churning within earth. This is what we call the core of alchemical theory. Physical reality was a ratio of substances and effects of exhalations. He had the idea that any substance can be converted into any other substance. They are all fundamentally united what we call usia as substance. If unintelligent nature performed this process deep within the realms of the earth by final causation as Aristotle thought. It was

an active intelligence that reproduce or even improve the process through artifice or through art itself just like the medieval alchemist. There was a task to reduce a substance back into the neutral substrate the medieval materia prima and then finally recompose it to a more noble form and during a point in antiquity by Maria the Jewish and Zosimosa Panoplis around 300 of the Common Era and the context of Alexandrian. The substrate already had the powers to convert baser metals to noble ones through artifice. The earliest reference to the substance as a stone is in zooms of panopolic where he decides to call a stone that was not considered to be a stone. The process of transmutation encoded like the formula of the crab and the chrysopoeia. Literally, we are talking about the gold making of Cleopatra merges both philosophical elements. This can be seen as the tail-eating snake or bodies or the saying that "All is One" along with the apparatus for transmutation. We can also see it as a distillation apparatus. Theory and experimentation philosophy and empirical science combined.

Hellenistic science was absorbed into what was referred to as then the new Islamic World. Alchemical theory became a good deal more precise rather than vague elements. The alchemical folks at Alexandria informed by the esoteric priestcraft of the Egyptian metallurgists were based on the coloring of metals and especially working with gold. Islamic and alchemists have developed sulfurmercury theory of metals. The theory held that metals were a specific ratio of mercury and sulfur either being fused through complex processes within earth or mercury alone as a

fundamental substrate subjected to chemical processes which converted it to a fixed noble metal with sulfur being an impurity. The exact origins of sulfur-mercury were actually lost and were likely hidden. The strange Ismaily circles associated with The quasi-legendary Baghdadi alchemist and philosopher Janet Emmett hayam. It was probably very likely to emerge out of careful experimentation.

We can explore this by a mature process by which the philosopher's stone was going to be produced as it was articulated in late medieval European Latin Alchemy. Know that historical alchemists were universally chemically analyzing and transforming nature. They were interested to make gold. The concept of spiritual alchemy or alchemy as a psychological process was a 19th century and 20th century romantic version. The spiritual or psychological reinterpretations of alchemy can be interesting and then projecting them onto historical alchemy distorts our understanding what the alchemists were doing and why they thought they were doing it. They had rigorous theories supported by experimental observation. There were at least as many theories how to produce the philosopher's stone as there were alchemists. We know now that alchemical theory was not correct. It was not a correct theory of nature and producing the philosopher's stone or any elemental transmutation was considered to be both not possible. Alchemists were also considered to be careful experimentalists discovering the truths about nature. There was even literature about the philosopher's stone to contain some combination of experimental information along with a significant speculative theory. There will

never be a singular process by which the stone of the philosopher's would be produced.

The alchemists never succeeded at discovering it because it was not discoverable. Let us take a look at how the philosopher's stone was going to be produced in many alchemical theories. The process had about five stages where the earlier stages were grounded in experimental evidence and the later which move and drift in pure alchemical speculation. The first stages had to do with preparing the base substances that were required for all alchemical experimentation scubas gold, silver, sulfur, mercury, and like the paracelsian school of salt or other chemical that were used in the process of transmutation. The preparation of metals was instructive for alchemical process and also for alchemical theory. You need base metals to be transformed during the alchemical process and lead being both common and expensive as well as easy to work with and toxic. The common forms of lead to occur in nature was the mineral galena or blood sulfite. In order to extract the elemental lead, one has to roast it over a common fire. The smell as it heats up gives the odor of sulfur. Galena is actually lead sulfide. With liquid lead that looks like mercury before being quenched in water and fixed as lead. Alchemists who worked with better fire control realize that using the right furnace and the right crucible and if they increase the temperature of the galena, then the lead was found to vanish. They would oxidize to leave behind minute quantities of silver which is another element found in the mineral galena. Roasting lead gives it a mercury liquid appearance before giving sulfur off as a smell left behind by transmutation. The sulfur mercury

theory sounds strange, but it does comes out of experimental experience. This proves the production of mercury or quicksilver. Gold and silver were further prepared by copulation heating so the base metals oxidize and vaporize with the furnace and crucible with lead sulfur and antimony.

Liquid mercury was indispensable for alchemy being extracted from the stone cinnabar. If we were to roast powdered cinnabar, the mercury sulfide fumes were then fed by distillation into cold water. The gaseous mercury vapor would be condensed to a liquid in cold water where sulfurous gas being released during this chemical process. Further proof of the sulfur-mercury theory of the metals could also produce some amount of sulfuric acid or oil of vitriol as the alchemists had it to be a byproduct.

The extraction and purification of the fundamental alchemical elements have been used. Most of these alchemists practices overlap with the process of metallurgy, determining the quality of metals, mining, and minting of coins but had to be mastered by an alchemist. The primitive elements need to be fit for alchemical use. Taking their properties of being sophic or a philosophically purified form is what we can partition the world of metallurgy and assaying and alchemy. The process of making soffit gold and silver involves dissolving gold with aqua regia and silver with aqua fortis. The solutions would be subjected to evaporation before being crushed into a powder. We can see why alchemists call this to be gold powder sophic sulfur which is the reddish gold powder resembled elemental base sulfur.

The green lion devouring the sun where the lion represents the acid and the sun being gold. The green tint represents the process itself. Most of the gold they were working with contained a bit of copper which gives rise the greenish color and acid solution. Silver powder was thought to be sophic mercury and quicksilver was thought to be a sophic salt or the minstream or matrix. This was the materia prima in which things could be transformed into the stone of the philosophers which again convert base metals into noble ones.

There were still disagreements in alchemical theory. Some schools held that elemental sulfur and mercury were the working part of the process, whereas others held that sophic sulfur and mercury along with quicksilver were part of it. Others held that mercury or quicksilver were part of the process. Sulfur was always thought of as an impurity to be driven off which left the stone behind. The Philosopher's Stone was the process of driving off the sulfur. Many of the sulfides required driving off the sulfur during the process of purification. Others felt there would be a wide range of substances to be introduced into the matrix for the production of the stone.

Continuing the process there was some ratio of sophic sulfur-mercury or quicksilver along with other potential substances would be sealed into a thick clear vessel. This thick clear vessel was considered to be the Hermetic egg. The vessel was also called the eluded or the vase of the philosophers along with other names. It was also shaped like a tear drop. The egg would be sealed from the outside world. This was what we call hermetically sealed. There were other shapes like pelican designs that could be used to proximate elements

of the great work. There have been alchemic texts that give a range of volume for the hermetic vessel but the glass must be quite thick. It also needs to be clear and withstand the pressure that build up within it. It was imagined not to be large since once the stone of the philosophers gets produced within it. Even a small amount converts massive amounts of base metals into noble metals. The hermetic egg gets subjected to fires or various intensities. It undergoes processes in a specific order and changes colors in a specific order as a stone comes into being in the matrix and the womb of the hermetic egg.

The careful control of both heat and fire was common during the days of alchemy. There were no standards to denote temperature that existed at this time, but there were a range of colorful conventions. It used to be called the heat of a nesting hen which was called the Egyptian fire. This was perceived as the heat of an Egyptian summer day cooking fire. As for the melting point of lead which was about 621.5 degrees Fahrenheit or 327.5 degrees Celsius. Celestial heat which comes from the stellar rays of astrological objects were considered to be the heat which was given off by a water bath.

What kind of heat and when in the process was endlessly debated by the alchemist, but if you did heat things properly in a proper way it can botch the whole process and of course overheating would destroy the transmutation process. The processes that give rise to the stone of the philosophers as described the order for which they occur has constantly been debated. Some of the alchemists argued about a singular process in the transformation to the stone where

other alchemists have argued about four processes. Paracelsus argued for just seven, whereas others argued as 14 though most tend to land on 12 which corresponds to the sign of the zodiac.

Calcination was the process of reducing the components that enter into the hermetic egg to powder through roasting. This process also results in what we call fixing the metals which makes them less fusible and also less liquid. Conjulation will be the introduction of sophic mercury and sophic silver with some quicksilver in the egg before sealing it depending on the school. The phase is illustrated as marital conjunction or chemical wedding. This was the most famous aspect of the alchemical process and art and also in jungian psychology. The great work of alchemy can really be said to have begun. It was the conjunction that gives rise to the work.

Solution and dissolution was the same thing from a fixed substance into a liquid one by using things such as acids or philosophical waters as the alchemists. This compound was mixed with revivification in which dead or fixed elements might be again made such as extracting liquid mercury from cinnabar. The processes of digestion or fermentation was thought to be a rare substance which could penetrate into a denser one through invisible small pores in the substance. This process can be helped with sedation, serration, or inhibition feed the fermentation or alchemical processes with various substance along the way.

A theory of the transmutation was the philosopher's stone penetrated into the pores of basic metals through fermentation and perfected them by sealing them in a type of manner. Lead really was not transformed into

gold, but it was perfected through the fermentation of the subtle power of the stone. The stone has the ability to heal the base metal which adds the medical dimension found in the field of alchemy as well as the medical powder of the stone to cure people and having the ability to reduce the signs of aging. This theory of the powers of the philosopher's stone was one of the few naturalistic theories for the power of the stone at all. Remember that this was just a citation theory. Distillation or the substances to separate through heating and going through condensation was a perennial favorite of the alchemists. Substances would be distilled a lot with the distillate reintroduced to the base in a process we call circulation or cohabitation. Using primitive reflux stills like the double pelican model were considered to be double lambics or multiple linked alambics. Sublimation was the formation of crystals towards the top of the hermetic egg or a ludic by heating and cooling the substance towards the top. The exaltation was thought to be the pure substrate of the substance being heated and these sublimates of essential salts. The alchemists would call them highly coveted substances. Both distillation and sublimation were diced in alchemical art and favorite procedures even though they return the distillate to the pot for red distillation did not have a chemical effect. It was often a waste of time from a chemical change. Separation was a term which covered a range of processes such as decanting and racking. For instance, recovering the oily sulfuric acid from the water used to cool mercurial vapor produced during the distillation of cinnabar. That was an example of racking for producing liquid mercury. Putrefaction or mortification were often induced through more heating would destroy the outer substance

while preserving the living metallic seeds within which then could be resurrected through many means. This was thought of as the most precarious throughout the alchemical process. It really did involve highly specific temperature control of philosophical fire. Too little with the seeds have been trapped in the substance and too much would be destroyed. The idea of metallic seeds has emerged from experimental experience. Model alchemical reconstruction of alchemical experiments show things look like trees inside. It reflects the hylozoic or the living matter theory of hermetic thought which was a difference with dead matter theory of 18th century materialism.

If you have ever seen the alchemical image of the 12 keys of Basil Valentine, the Splendor Solis, or the mutus libra, then it was likely you have seen a depiction of some alchemical substances undergoing one of these or many of the processes of. But there were still two more important steps which include multiplication and projection. All the processes going on in the alchemical egg were thought to be changes in the color of the substances going on there within the hermetic egg. The changes were the only indicators of the process proceeding correctly, and if they did not appear in the correct order, then something was wrong. The specific colors and their odors was debatable. But during the days of early alchemy, there were four such colors such as black then white then yellow and finally red. Another important color would be the peacock's tale which occurred between black and white as well as the green lion and if the colors were not in the correct order such as red coming first, it was thought that overheating would happen and the experiment would fail. In all chemical art the colors take on a wide range of representations with flowers and birds such as the raven, the swarm the peacock, the phoenix, etc. How long this process would take would be the subject of enormous debate. But the alchemists all agreed that patience and faithful was in order. Some of them would have the process taking as little as seven days that mirrors the process of the divine creation and a day of rest. Others took an entire year, whereas other track more specific things for the procession of the zodiac for the beginning and the completion of the process. There were even some thought the stone would only transmute silver which would take less time to produce than one that also transmitted gold. We are talking about five months for the silver stone and between seven and nine for the gold one whereas others have three, seven, or twelve years.

Others would take a single season for how long it could take. A popular unit of time would be the philosophical month of 40 days with each color change that proceeded through one of those philosophical months about 160 days. The processes took time and all chemical texts urge patients to be a sublime virtue. While the middle parts were widely debated, the final parts of the alchemical parts were agreed upon. Towards the end as the substance became a bright shade of red, the stone of the philosophers began to emerge in the first week infantile form. It was thought it needed to be fed just like a baby with mercury to make the stone stronger through the process of sublimation. In fact, this process of multiplication concentrates the power of the stone and when it was complete, the stone has the power to translate huge enormous quantities of base metals with very small amounts of the

philosopher's stone. The stone would look like a kind of wax-like substance or a red gem or perhaps like a ruby, but it was commonly a red powder that was heavy and having a wonderful aroma. The powder would be projected as the final state of the alchemical process by casting small amounts of it either on molten lead or quicksilver or mercury where the base substance transmuting into either silver or some white powder to be a lesser philosopher's stone or the production of gold via the red stone or red powder.

Remember we talked about the appearance of sophic sulfur prepared from the gold at the start of this alchemical process. The silver or gold was mentioned to be more purer than that aside from the mines art. The stone of the philosophers had the ability to cure disease and have the ability to reverse old age. It could convert common stones to precious gems. It could even cause crops to grow in barren lands. It even had the ability to speak the languages of beasts and birds, and it had the ability to have an individual communicate with the angels. You can only taste it to be subtle. Like legends over time, the stone came to have significant power beyond the ability to create worldly wealth.

We can think of the philosopher's stone as a concept which interacts with early experimental science with philosophical speculation with theological concerns about alchemy and having the ability to perfect nature. Fields such as medical theories, philosophy, spirituality, and psychology all intertwined together describe the philosopher's stone. The philosopher stone was taken as an interior state of one psychology or a spiritual being. These powers of the stone were captured in hundreds of

names in alchemical literature such as the shadow of the sun, the azatha paracelsus, the sophic hydrolith, the Basilisk of the philosophers, the blood of salamander, the universal quintessence, or substantial exuberantia. It was considered to be a stone but not really a stone as Zosemus had it along a millennia and longer. The stone continues to be the subject of fascination and even the object of more alchemists who might still be out there.

> ***"Of all Elixirs, gold is supreme and the most important for us...gold can keep the body indestructible.... drinking gold will cure all illnesses, it renews and restores." Paracelsus (1493-1541 AD).***

There were few objects that had as much interest as The Philosopher's Stone which was a legendary substance that transforms base metals into gold. During the Middle Ages and also throughout the late seventeenth century, one of the main objects for alchemists was to find this philosopher's stone. It comes as a surprise that many of the Western world's greatest minds including Sir Isaac Newton, John Dee, and Robert Boyle searched for the philosopher's stone.

We may ask what is the mythical stone and why did everyone want their hands on it? The earliest written manuscript mentioned of the philosopher's stone dates back from 300 AD to ancient Greece and was found to be in Cheirokmeta by Zosimos of Panapolis. Some alchemical writers claim that the history goes back further to the time of Adam who ends up acquired knowledge of the stone from god. It was claimed that there was knowledge of the stone that gave early biblical patriarchs unusually longer lives. The

idea of the philosopher's stone was taken hold and then continued to advance. This concept then developed in the medieval Islamic world through works of outstanding alchemists such as Jabir bin Hayyan. It was his work which laid the foundation for later alchemical practices in the Islamic world and also in Europe. He introduced essential processes and different types of alchemy who wrote over 300 texts on this subject. The knowledge permeated into medieval Europe where this idea of the philosopher's stone became integral to Western alchemical practices and various theories. European alchemy had important figures such as Albertus Magnus, Roger Bacon, and the Swiss physician Paracelsus played crucial roles shaping the lore of the philosopher's stone.

Paracelsus felt their existed an undiscovered element called alkahest from all other elements such as earth, fire, water, and air. He felt the unknown element was the famous philosopher's stone. The symbol for the philosopher's stone features each of the symbols of four elements being fused together. We will be talking about "The Golden Tract" in the next chapter. In this tract, there was an unknown German author that expounds this theory of the stone as the fifth element. He writes

> ***"Spirits, and soul, and body, and four elements: the fifth which they furnish is the philosopher's stone."***

Alchemists felt everything was composed of spirit, soul, and body. The four elements were considered to be traditional, but the fifth element was considered to be the philosopher's stone which symbolized a transcendent essence beyond the physical realm. They felt that the philosopher's stone contained the four classical elements

such as earth, air, fire, and water combined and symbolized the ultimate balance and perfection.

Even this concept of the philosopher's stone has the equivalents in eastern philosophies. In Buddhism and Hinduism, this concept is called the Chintamani which depicts a wishfulfilling jewel. This embodies the transformation of physical substances just like the Western alchemical tradition, but symbolizes profound spiritual transformation and enlightenment. There were many different descriptions what the philosophers stone looks like. Based on alchemical texts, it could be either white or red. The white variety turns base metals to silver and there becomes a less mature version of the red stone. It was the red stone that makes gold. It was described by some as being orange or saffron colored in the solid form, and then red when ground to powder. It was heavier than gold, soluble in any liquid, and also able to withstand fire. Alchemists believe the philosopher's stone could be created through the alchemical method known as the magnum opus or what we call "The Great Work". The secret and complex process was divided into four main stages that represents by color: nigredo (blackening) which symbolizes decay and purification, albedo (whitening) which indicates purification and the washing of impurities away, citrinitas (yellowing) which is a stage of enlightenment and drawing true knowledge, and rubbed (reddening) which was the final phase representing achievement of the highest state of enlightenment and creation of the philosopher's stone.

There were many different methods describing how the philosopher's stone can be created. There was a seventeenth century mystical text called “The Mutus Liber” (or Mute Book) which appeared to be a symbolic instruction manual creating the stone. It has fifteen different illustrations and have no words. Even Sir Isaac Newton has decided to leave behind a procedure to make the philosopher’s stone which appeared to be a copy of another well-known alchemist’s text. It describes how a philosophical mercury substance was a key ingredient in making the stone and could then be broken and reassembled to make different metals. Once the stone was made, it would have many powerful properties. The philosopher’s stone turned base metals to gold, but it was thought it could heal any type of disease and grant immortality. It was considered to be a key ingredient in the mythical potion “The Elixir of Life”. No wonder a lot of people were after this stone even from kings to the paupers.

The philosopher's son was about the transmutation of base metals into gold, but it also symbolized a spiritual journey towards enlightenment and a state of perfection. This was also reflected in key alchemical texts such as "The Emerald Tablet". We will be talking about this tablet in a later chapter. This tablet was attributed to the legendary Hermes Trismegistus which gives the idea of the unity of the cosmos and the individual.

The philosopher's stone over time transcended its origins to become symbolized as spiritual transformation, wisdom, and the pursuit of perfection. The symbolism of the philosopher's stone can be understood in terms of spiritual transformation and enlightenment. The stone itself is a physical substance and a metaphor of spiritual wisdom and enlightenment. Turning base metals into gold was analogous to achieving spiritual awakening and transformation. The philosopher's stone symbolized reconciliation of opposites which was a central theme in alchemy. This even includes the combining of masculine and feminine, earthly and divine, or even the material and the spiritual. It also represents the achievement of both harmony and balance. As mentioned before, the quest for the philosopher's stone was referred to the magnum opus or "The Great Work" in alchemy. This was a journey which encompasses physical, moral, and spiritual purification. The stages for the process were encoded in symbols and allegories.

The philosophers stone was associated with "The Elixir of Life" which was a mythical potion granting immortality. This symbolized the human quest for longevity, health, and overcoming limits of physical existence. Remarkably, the stone also symbolized the achievement of the perfected state which was both on personal levels in perfect wisdom or enlightenment and a cosmic level as in the perfection of the entire universe. The philosopher's stone represented the goal of alchemical practice and also considered to be the highest form of mystical knowledge. Possessing the stone was thought to give the meaning of profound understanding of God, nature, and also to oneself.

There were many influential and intelligent people who searched for the stone for many decades. There was no proof of anyone who has discovered it. There were some who felt they have found it such as Nicholas Flamel who was an ordinary craftsman who also became extremely wealthy and felt he found the philosopher's stone. But there was no evidence to back his own claims. Instead, alchemists found other things during the process of chasing the philosopher's stone. They discovered nitric acid, sulfuric acid, and acetic acid and even the creation of alcohol. They also discovered new salts and new substances such as antimony, arsenic, zinc, bismuth, and other nonmetals such as sulfur and carbon. They were the first system who created the chemical elements in the world. Alchemy also opened doors to more scientific methods of study and research and opened the doors to the world of chemistry as we know it today. As the fields of chemistry rose, alchemy started to decline. The concept of the philosopher's stone paralleled other mythical objects who came across cultures like the Holy Grail or The Elixir of Life. These comparisons highlighted the human fascination for achieving eternal life, wisdom, and transformation. It was a theme that transcended geographical and cultural boundaries. Surprisingly, there are people out there who still search for the philosopher's stone and think they can create the most powerful substances only if they get the chemicals right, but in reality, it is highly unlikely that this will ever be achieved physically. Alchemy carried out principles that we now know are not scientifically valid. Gold can be created from other elements but nuclear reactions would not to be involved.

> ***"The alchemists were indeed correct that lead could be turned into gold-even if they were dead wrong about how it could be done."***

During modern times, the philosopher's stone holds significance especially in esoteric and spiritual circles. In Jungian philosophy, the philosopher's stone was seen as the symbolic transformation of

oneself. The psychoanalyst Carl Jung felt the experiments and mysteries language were about personal development and achieving a balanced complete oneself. He used it as a type of metaphor for individuation and saw the philosopher's stone as a quest for self-actualization. The philosopher's stone was interpreted in symbolic form to represent personal transformation and enlightenment. The modern interpretation changes from focusing on the object itself to a metaphorical journey of self-discovery and also spiritual growth which reflects the evolving nature of alchemical thought from the physical pursuit to the quest for inner wisdom. We can find the meanings of the philosopher's stone in "The Golden Tractate".

Chapter 11

THE GOLDEN TRACTATE

The Golden Tractate of Hermes Tris Magistus was the alchemical manuscript of the philosopher's stone. Hermes through his long years has not ceased to experiment not spared his labor of mind and science and art to obtain by the inspiration of the living god who judged to open them. It was the servant who has given rational creatures the power of thinking and judging the right to forsake none or to give any occasion to despair for himself I have never discovered the matter to anyone who has not been fear of the day of judgement and the perdition of soul to conceal it. It was a debt for which he had a desire to discharge to the faithful as a father of the faithful. He felt the sons of wisdom that the knowledge of the four elements were ancient philosophers were not sought after through patience to be discovered according to the causes and the occult operation. But the operation was a cult since nothing was done except the matter to be decompounded and since it was not perfected unless the colors would be passed and then accomplished. The division was made upon water by ancient philosophers who separated it into four substances which included one into two and three and into one the third part which was the color as it was a coagulated moisture, but the second and third waters were the weight of the Y's who took the humidity or moisture and ounce and a half and/or the southern redness which was considered to be the soul of gold. A fourth part was a half an ounce of the citrine or like half an once of aura pigment. Half an ounce which were eight that was three ounces and by knowing the vine of the wise was drawn forth and three, but the wine was not perfected until the length of thirty was accomplished. One should understand the operation and decoction to lessen the matter but the tincture augments since the luna in fifteen days becomes diminished and in the third she was augmented. This was the beginning and the end. What was hidden was within fixed and had it either in earth or sea to keep to keep the argent vive which was prepared in the innermost chamber for which it was coagulated for which mercury was separated from residual earth. Those who hear his words let them search into them and also justify no evildoer but to benefit for the good. He has discovered all of the things that were hidden concerning the knowledge and disclosed the greatest of all secrets including the intellectual science. He wants you to know the children of wisdom who concern the report that the vultures stand upon the mountain to cry out with a loud voice. Hermes states that he was the white of the black and the red of the white and the citrine of the red. He speaks of the truth and know that the chief principle was the crow which was the blackness of the night and clearness of the day and also flies without wings from bitterness that exist in the throat the tincture was taken. The red goes forth from the body and from the back it was taken as thin water.

He wants you to understand accept the gift of God which was hidden from the thoughtless world. In the caverns of metals, there was a hidden stone

that was venerable splendid in color in color and a mind sublime and an open sea. He declares to give thanks to the God who teaches the knowledge and in return recompense as the grateful to put matter into a moist fire and also to cause it to boil to heat which then destroys an incombustible nature until the radix appears the redness and the light parts of only a third remains of science.

The philosophers were said to be envious and not that they grudge the truth to religious or men or to the wise but also to fools who are without self-control. They are made powerful and are able to penetrate sinful things for the philosophers were made accountable to God and evil men who were not worthy of the wisdom. Know that the matter we call the stone but was named the feminine of the magnesia or the hen or the white spittle or the volatile milk or the incombustible oil so it may be hidden from ignorant who were deficient and goodness and self-control which have signified to the wise by one.

The philosopher's stone includes and conserves the fire and the heavenly bird to the latest moment of an exit, but he deprecated all sons of philosophy to whom the great gift of knowledge which should be bestowed if any undervalues or divulge the power to the ignorant or being unfit for the knowledge of the secret. He mentions that he has received nothing from any who have not returned and that which has been given to me nor has failed to honor him even the highest confidence would the concealed stone of many colors that was born and brought forth in one color know this and conceal it by the almighty who favors the greatest diseases being escaped and every

sorrow distress as well as the evil and hurtful thing to depart. This leads to the darkness into light from this desert wilderness to a secure habitation and then from poverty and straits to the free and ample fortune.

According to the second section, those who admonish the fear of God and whom the strength of the undertaking and the bond of meditation to unloose and hold not to be a fool which lies hold of his instructions and meditate upon them. The heart should be fitted to conceive as the author himself he teaches to lie too cold to any nature that it would be hot and it should not hurt in a manner who is rational shuts himself within from the threshold of ignorance blessed and he should be deceived.

He states that we should take the flying bird and drown it flying and then dividing and separating it from pollutions which hold it in death draw forth and repel it from itself and that may live and answer by not flying away into the regions but by bearing to fly for it to deliver out of its prison and according to reason and according to days he teaches it will become a companion and become to be an honored Lord extracted from RACI its shadow and from light its obscurity by which clouds hang over and keep light away by construction and fiery redness it was burned to take his son the redness corrupted with water as a live coal that holds the fire which should withdraw until the redness made pure and then it will associate by whom it was cherished and in whom it rests return then. The coal is extinct in life upon the water for thirty daises he notes a crowned king who was resting over the fountain and drawing dense the aura pigment dry which does not have moisture and have made the heart of the hearers hoping to rejoice in the eyes beholding in anticipation of that which possesses.

The water was first in the air than the earth restore it as well as the superiors by proper windings and not altering it to the former spirit fathered in redness which should be carefully conjoined that the fatness of our earth was sulphur the aura pigment CRT and which sulphur were pigments such as sulfur. Some are considered to be more vile than others for which a diversity of which kind was the fat of gluey matters such as hair nails, poof's and sulfur and brain which was our pigment of likekind, and the lions and cats clause which was the fat of white bodies and the fat of two oriental quicksilvers sulfurs hunted and retained by the bodies. He states the sulfur is held by the conjunction of tincture oils but fly away in the body contained which was a conjunction of fugitives only with sulfurs and abdomens bodies that hold and detain the fugitive ends. The position sought after by philosophers was but one in an egg and this in the hen's egg was much less to be found much of divine wisdom in a hen's egg which should not be distinguished. Our composition was from the four elements adapted and composed in the hen's egg to be the greatest help with respect to proximity and relationship of a matter in nature for there was spirituality and conjunction of elements and an earth which was golden in tincture but the son enquiring or the sulphur which fit our work whether they are celestial or terrestrial to whom the father answers them as heavenly and some of the earth imagined the heart in the superiors to be heaven and in the inferiors earth it was not so masculine as the seven of the feminine and the feminine was the earth of the masculine.

The Sun asks father which of these was more worthy than the other whether it was the heaven or the earth. Hermes replies both were in need of help one of the other for demanding a medium but the wise man governs all mankind but ordinary men were better for them since every nature delights in society and we can find it in the life of wisdom that we were equals conjoined but what rejoins the sun is the mean between them in nature. There are three from the two in the beginning the middle, the middle, and the end which was the first needful water and then the oily tincture and last the faces of earth which remained below but the dragon inhabits these and his houses of darkness and blackness in them and by them he then ascends into the air from rising which was their heaven but the fume remains in them. They were not immortal. They should take away the vapor from the water and the blackness from oil texture and death from faces and by dissolution should possess a triumphant reward even in the possessors live that the temperate undulant which was the fire and was also the medium between phases and water and were the water and the sulphurs between fire and oil and the sulfur there was such a close proximity that even the fire burns as it does the sulfur and then the sciences of the world comprehended this was my hidden wisdom and the learning of the art that consisted of wonderful hidden elements which adopt and discover and then complete. It behooves him that he was introduced to the hidden wisdom to free himself from hidden usurpations and be sound reason ready to hand to help those of a serene countenance diligent to save and be a patient guardian of arcane secrets of philosophy and to understand how to mortify and induce generation to

liven the spirit and introduce light until they fight each other and then grow white and free from defilements rising as if it were from blackness and darkness and know nothing nor cannot perform anything. But if one knows of a great dignity so even the kings should reverence the secrets son it to be concealed from the vulgar and profane world. Understand that the stone was from many things and various colors and were composed of from four elements which we should divide into pieces and then segregate in the veins and mortify the same by the proper nature which was also in it to preserve the water and fire dwelling which was from the four elements and their waters that contain the water was not water in its true form but also fire which contained a pure vessel the ascending waters should fly away from the bodies or by means they should be made fixed whole blessed watery form which dissolves the elements with the watery soul to possess ourselves of a sulfurous form and to also mingle the same for when by the power of the water the composition was dissolved was the key of the restoration and then the darkness and death should fly away from them and then wisdom proceeds onwards to the fulfillment of the law.

According to the third section, the philosophers bind their matter with a strong chain that contends with fire because the spirits in the wash bodies desire to dwell and rejoice in the habitations themselves and in habit there and the bodies hold them nor can they be separated any more of dead elements were revived the composed bodies and being altered by a wonderful process. They were made permanent water to reform and create the royal elements and a government obtain the tincture fine to be distressed

of out most precious stone. It was made worthy that would be vile and behooves us to mortify together and to be venerated. The stone comes to light with light being generated and then generating and bringing forth the black clouds or darkness which was the mother of all things but the crowned king is married to our red daughter and in a gentle fire not be so hurtful she conceived an excellent and supernatural sun which was permanent life she fed with subtle heat that he lives at length in the fire but when they were sent forth the fire upon the foliated sulfur was where the boundary of heart enter in above, and it was washed in the same an purified matter extracted was it that he transformed and the tincture by the help of the fire remained red as it were flesh but the son the kind begotten took his tincture from the fire and death even and darkness and the waters which flee away the dragon and then blocks the sum beams which dart the crevices and the dead son who lives the king come forth from the fire and rejoins with his spouse. The occult treasure were laid open and the Virgin;s milk was whitened. The sun vivified and became a warrior in the fire of tincture for this son was himself in the treasury bearing the philosophic matter approach wisdom and rejoice and now rejoice together for the reign of death was finished and the sun rules and now he was invested with the red garment and the scarlet color being put on.

According to the fourth section, understand the wisdom what the stone declares to protect me and he will protect the increase of my strength that may help the soul and the most inward and secretly in me my own luna and also my light exceeding every light and my good things were better than

the other good things. He gives freely and rewards the intelligent with both joy and gladness filled with glory riches and delights and seeks to understand and possess divine things which the philosophers had concealed and which was written with seven letters of alpha following the soul and follows the book which was willing that he should have dominion observe the art and join the son and even the daughter of the water which Jupiter had a hidden secret auditor understand. Consider the most accurate investigation which can be demonstrated to the whole matter to be only a thing but who understands the true investigation and inquires into the matter. It was not from man nor from anything like him and if there was a creature who joins with one another species which was brought forth a neutral form either from light nor darkness of his nature and if his mettle was not dried all bodies desire them and wipe away their rust even if their substance was extracted.

Nothing was better or more venerable than his brother being conjoined but the king who was the ruler testifying he was crowned and he was adorned with royal diadem. He holds the royal garment and brings joy and gladness of heart were being chained since this has caused the substances to lay hold of and to rest within the arms and breasts of mother and to fasten upon the substance making which was invisible to become visible and the occult matter appeared and everything which the philosophers have hidden was generated. These words are being mentioned and understand them and also keep them and meditate as well as seek for nothing more man from the beginning is generated of nature and whose inward substance was fleshly and was not from anything else. Mediate on these plain things and reject the superfluous the philosopher tree has made from the citrine which was extracted out of the red root and then from nothing else and if it be citrine and nothing else wisdom was there, and it was not gotten by the care nor has it been freed from redness. He has circumscribed nothing of there was understanding there but few things can be unopened to sons of wisdom who turned the brain-body with an exceeding great fire, and it will yield what you desire and observe you make that which was volatile. So it cannot fly and by means the fire as it were a fiery flame and that which in the heat of a boiling fire was corrupted and know the art of this permanent water was our brass and the colorings of its tincture and blackness was then changed into true red. By the help of God, there was nothing but the truth that which was destroyed and then becomes renovated and hence the corruption was made manifest in the matter to be renewed and hence will appear on either side of a signal art.

According to the fifth section, the tractate states that the son which was born of the crow was the beginning of the art behold that it was obscured a matter treated by circumlocution deprived light and joined the nearest and farthest to those things and boil them in which comes from the horse's belly for seven, fourteen, or twenty-one days then the dragon will eat his own wings and destroy what was done to be put into a fiery furnace and observe none of the spirit may escape and know the periods of the earth were in water and which let it be as long as until the same upon it the matter was melted and burned and triggered the most sharp vinegar until it becomes obscured.

This lives in the putrefaction and let the dark clouds which were in it before was killed to be converted into its own body and let this process be repeated as described it again die and then lives on the death and life we work with the spirits for as it dies by taking away of the spirit it lives in the return and has revived and rejoiced there to be arrived at the knowledge that has been searching for was made in the affirmation. I have related to the joyful signs which fixed the body but these things and how they attained the knowledge of the secret was given by the ancestors in figure and types they were dead opened the riddle and the book of knowledge has revealed hidden things uncovered and brought together the scattered truths within the boundary and have conjoined in many different forms. Associate with the spirit and take it as a gift of god.

According to the sixth section, it behooves one to give thank to God who has bestowed of the bounty to the wise and who delivered us from misery and poverty. I have proven with the fullness of the substance and his wonders and pray God that we live and we may come to him. Remove the sons of science the undue ins which we can extract from fats and bones which were written in the books of the fathers but concerned the ointments which contained tinctures that coagulate the fugitive and adorn the sulphur and also behooves one to explain their disposition more at large and to unveil the form which was buried and hidden but also dwells in the body of fire in trees and stones. It behooves to extract without burning and know that the heaven was to be joined immediately with earth but the form was in the middle nature between the heaven and earth which was the water but the water holds of all the first

place which goes forth from this stone but the second was gold and the third was gold to mean which was more noble than water and the faces. These represent the smoke which was the blackness and the death. It behooves us to dry away the vapor from water and then to expel the blackness from the unjoined and death from the feces and this by dissolution means we attain the highest philosophy and secret of all hidden things.

According to the seventh section, know the sons of science there were seven bodies of gold which was the first and most of the king of them and at the head which neither the earth can corrupt nor fire devastate nor the water change its complexion becomes equalized and nature regulated with respect to heat, cold, and moisture nor was there anything in which was superfluous and therefore the philosophers Dubuis magnify themselves. It was like saying that the gold in relation of other bodies as the sun among the stars more splendid in light and as by the powers of god that every vegetable and all fruits of the earth were perfected so gold by this same power sustained for as dull without a ferment cannot be fermented so when one sublime was the body and purifies it by separating the uncleanness from it which will then conjoin and mix the together and put in the ferment confetti the earth and water, then matters ferment and how this ferment in this case has changed the former natures to another thing observe that there was no ferment. Deferment whitens the confection and hinders it from turning and holding the tincture which should fly and rejoice the bodies and also makes them intimately join and to enter one another. This was the key of the philosophers and the end of the work and by the science bodies

and makes them to join and enter into one another, and this was the key of the philosopher's and the end of the work and by science bodies were military rated and operation of them by God. It was God who assists and being consummate but through negligence and a false opinion of matter then the operation may be perverted as a mass of leaven that grows corrupt or milk turned with rennet for cheese and the musk for aromatics will be the sure color of the golden matter for the red and the nature was not sweetness. They are made and come as a mixer and we make the enamel which we already without the kings seal that we tinged the clay and in that there was a set of color of heaven which augments the sight of them and we see the stone as the most precious gold without spots being tempered which neither fire not air nor water nor earth was able to corrupt for it was the universal ferment that rectifies all things in a medium of composition who complexion was yellow and the true citrine color. The gold of the wise boiled and digested with a fiery water makes sixth year for the gold of the wise to be more heavy than lead which in a temperate composition was a ferment year. This gives contrary wise out of intemperate composition, and it was the confusion of the whole for the work began from the vegetable which was next from the animal as in a hen's egg in which the great help and earth was gold of all which we come as a firmament mixer.

The Golden Tractate as just mentioned was complex, but nevertheless a lot of things were not understood at this time. We are going to see up next that "The Emerald Tablet" continue to contain much more esoteric and mysterious language during the days of alchemy.

"The Highest Knowledge is..."

Chapter 12

THE EMERALD TABLET

Just before the untimely death around the age of 37, Hermann Rorschach was working on developing a psychological test for abstract images or ink blots that were used to reveal the unknown or hidden dimension of what we call the unconscious mind. There were original ten cards from black and white blots to colorful abstractions that were thought to provide a secret passageway into the emotional and psychological states of patient revealing to the analyst patterns of disordered emotions and thought patterns. We can think of this test as having gone to great fame and use which has also been derided as a pseudoscience and would be no different than cold reading. Studying the hermetic and occult literature in certain texts seems like to follow The Rorschach Test. The obscurity, brevity, and even the seriousness indeed invites enormous intellectual and spiritual attention and into the obscurity, brevity, and seriousness of projects for their own philosophical and religious concerns, commitments, biases, and desires in hermetic hall of mirrors. The own image was reflected back to them though because of the distortion they won't recognize as themselves.

Many texts immediately can be thought of as having the power as the book of formation. The mysterious letters towards the beginning of some surahs of the Quran, which was the channeled text of the book, revealed to Aleister Crowley during April 1904 though this effect has been considered nowhere to be present to me than in the famed Emerald Tablet of Hermes Tres Maguestus. In fact, the text has only a dozen or so line to contain all of the wisdom of hermetic philosophy from alchemy to cosmology to even the creation of the transformational philosopher's atom towards spiritual transformation. It is still interesting to note the text contains a volume length commentaries for many centuries and there were interpretations as there were versions of the text itself. The Emerald Tablet may have the greatest and poorly understood of text in the hermetic and occult philosophies.

Know The Emerald Tablet of Hermes Tres Maguestus would be considered to be the most famous, mysterious, and perplexing statement of hermetic philosophy. The religious philosophy was ascribed to the legendary Hermes Tres Maguestus could be analyzed into two artificial groupings.

The first was the so-called philosophical hermetic which were the texts that have survived from late antiquity and also describes a calcific religious philosophy. The initiate becomes reunited with the divine or mind through a process of spiritual education of ritual observance and seem alterations of consciousness.

The second group of the hermetic texts seems to have got less attention and are difficult to access even today. These were technical hermetic texts that cover a wide range of topics especially in the fields of astrology, alchemy, and even magic. The texts were preserved in

confused forms and were quite popular through the late classical period and then into the Islamic period and through the christian middle ages and then towards the modern period. It was this emerald table that sits directly between the two groups, and it seems to have a religious philosophical dimension and the cosmogenesis aspect has been understood as alchemical. It looks like we have the first enigma about the text. You may already be curious what the emerald tablet is as we will now explore the origins of the emerald tablet.

The oldest example of this text appeared in the arabic book "Kitab Sir Akalika" which was the secret of creation as well as the art of nature ascribed to the first century miracle worker as Polonius of Tiana or Baluns. He was known in the arabic world. The work was an encyclopedia of many topics including natural philosophy especially an early mention of the sulfur mercury theory of metals. This arabic text dates to around the early ninth century of a common era but the text has been claimed to be a translation from Syriac of an original text found in Greek. There was still disagreement of this question. A previous generation of scholarship accepted the textual history with more recent scholarships which the work as being composed of some amount of earlier material though largely being composed originally in arabic. It was probably in esoteric Ismaili circles for which Hermes Tres Maguistus was a prophetic figure for whom revealed wisdom took the form for scientific literacy.

The question was whether or not the emerald tablet was found at the end of the text from a remote period in antiquity or was in originally composed in arabic. Two other versions of the

emerald tablet were known in arabic which was one from about a century later. This one was preserved in the haberian alchemical corpus and seemed to be truncated and even corrupt though shorter variants of the text were sometimes indicator of a text but it was not possible to know. The other comes from the tenth century and comes from the pseudoaristotelean text "Kitab Asurar" which was known in the latin west as "The Secret Secretorum." We still doe not know if the text was more ancient than the appearance in the arabic heretic philosophy during the ninth century. There have been no reference to this text including late classical sources and greek or the syriac version had neither been found.

The text was so brief and enigmatic that evidence of translation from one language to another language does not appear. As an example, if we could find an example of hellenism or even a transcribed greek word would be evidence of greater antiquity, but still none of it jump out there could be an except to that found in both the seer al-khalika version. There was an important line that was missing from the jaberian text with an important technical term which was the arabic word for talisman. It was the arabic word talisman found in the line reading something like father of talisman's protector of wonders to be all powerful. This cuts in two different directions. The first direction was that this line has a survival of the greek word telesma which by the late classical Byzantine period had taken over religious or occult significance. It was something like the modern word talisman which meant charm if the word may be evidence of a late classical greek substratum for the emerald tablet. The text reads something that would be found in the alchemical

literature of people at the time such as Pseudodemocritus Zosimos of Panopoulos Stefanos of Alexandria. The problem of the theory was one there was an overall lack of evidence for the emerald tablet and antiquary and two the arabic word talisman was very well attested in exactly the period we were talking about. There was a period we were talking about, and there was an entire book on talismans written by Habit Ibn Kura that survives in latin but was almost exactly contemporary wit the Kitab Sir alkaline. We find that the earliest attestation of the emerald tablet such as a talisman reference that makes a lot of sense in the arabic hermetic era of the ninth century. It was still not clear whether the emerald tablet appeared in late classical greek or as a novel ninth century arabic use. It might be possible that the emerald tablet was a late antique text which was translated from greek to syriac to arabic which have gained the attention of folks in the esoteric Ismaili era. But the problem was we do not have evidence of any of that. Yet we can still point to other more circumstantial notions.

As an example, there was an oblique reference on one of the stobayan fragments of Hermes Maguestus and have his teachings quote engraved on a tablet and on a obelisk and in the nakamati hermetic text the discourse on either and ninth centuries there was a fact of admonition to copy out his teachings on quote steels of turquoise in hieroglyphic characters. But these documents were not the teachings mentioned like those found in the emerald tablet. There was no surviving hermetic texts that resemble the emerald tablet so perhaps it was possible the emerald tablet that predates the ninth century arabic.

The early history of this emerald tablet was rather mysterious as the transmission of the text into Europe was much better understood. During the twelfth century, there came to be an explosion of translation work from arabic to latin. Three of those translators and scholars Hermann of Corinthia whose day may have the earliest mention of the emerald tablet and the discovery in European literature there was also an anonymous translator that may have been plato of Tivoli who probably made the first translation of the emerald tablet and to latin and also Hugo of Santana who rendered the citable seer al-kalika from arabic to latin which includes the emerald tablet.

There was an early anonymous version rendered by Plato of Tivoli that would go on to be used by many celebrities such as Albert the Great, the teacher of Thomas Aquinas, and Arnold of

Villanova who was one of the greatest alchemists during this time period.

There was another translation of this text which would appear during this time from the pseudoaristotelain seer al-asurar or the secret secretorium or the secret of the secrets which was first translated by John of Seville and then a more full edition by Philip of Trivoli during the early thirteenth century. THe secret secretary was perhaps the single most read book during the middle ages after the Bible. It was like an encyclopedia that covered topics from medicine, physiognomy, and astrology.

By the thirteenth century, it would come to include some alchemy though the Hebrew translation argues that alchemy was a nonsense false science and it was impossible to carry out transmutation. This Hebrew translation of the sacretum secretorium was the

basis for the russian translation of the text. The popularity of this emerald tablet may have to do most with the general popularity of the sacretum secretorum. None of these versions became a popular translation and edition of the emerald tablet during the middle ages until this very day. There was a popular version of the emerald tablet known as the vulgate version. It was then introduced as another anonymous translation of this text with extensive commentary to enter Europe known as the libra hermetic de alchemia from an unknown arabic original which tells us the arabic alchemists have been using this text and were perplexed by it and were then composing extensive commentaries to elucidate what the book actually meant. This was something that would continue into the European context.

The translator from arabic to latin did not really understand or there may already been a type of corruption of the arabic word talisman. The line we mentioned earlier got rendered as pater Omnis tiles me tortillas with the arabic word talisman being mutilated in the translation into telesmi. Alchemists tried to grasp the hidden depths of this mysterious term and then poured a good deal of lab work and commentary onto its mysteries, and it was basically a translation error and not the arabic original such as the father of talismans.

There were other translation problems of the text. There were other texts that could not sort out if pater thesaurus miraculorum in the secret secretorium translation which was the father of the storehouse of wonders or in Hugo's rendering it was prestigious file which was the sons of talisman fathers or son or which one of these.

The problem we find is understanding the word abu in the arabic and mysterious use of the term talisman. The whole line was just a mess regardless of the version in the libra hermentis de luchemia which became the standard of vulgate version used through the entire medieval period and into the early modern period. If we were to study the emerald tablet it was probably this latin version or a translation into a vernacular language. Defective translations into latin from multiple different mysterious arabic versions ends up compromising the understanding of the text. To attempt an understanding of the emerald tablet there has to be a version and if you read latin you were still reading a translation and compromised the translation at that, and if you read arabic you are in better shape but then there were still three different versions to pick from all of which differs substantially from one another. There was a shorter one in the jiberian corpus that does not have the whole complicated talisman line and its brevity and corruption may even be an indicator of translation making it close to the text of the tablet. If we were going to interpret the text, then we have to pick some version in some language and go from there and an important question is on what rational grounds does one pick a version as more less authentic.

There were even European scholars who did not know the arabic texts into the early twentieth century which meant that all alchemists and all those hermetic philosophers from the translators of the twelve century to the scholars of the early twentieth century worked hard for nearly nine hundred years with many defective translations. Many of those people produced substantial commentaries

of this text to uncover mysteries that might likely had to do with garbled latin translations of a mysterious arabic text. We may ask what the emerald tablet say and what does it mean. Many version of this text were framed with a discovery legend where various figures from Sarah the wife of Avraham to Appolliunius of Tiana to Alexander the Great discovered an underground tomb which was thought to be the tomb of Hermes Tres Maguestus. The tomb was located near heron where there was a skeleton to be holding a tablet engraved on a green stone being interpreted to be the emerald which gives the name containing a brief text and being described as being written in an ancient language in some form of Aramaic and even Phoenician. The theme find many kinds of treasures and ancient tombs was popular in antiquity but because it happened and this also reached back to Plato's ring of gaijin's myth or in some sense survives even into Ridley Scott's famous scene with an engineer and alien. The text was considered to be very brief. The vulgate latin version was only 151 words and the shortest arabic text was only about 50 words. It seems to describe a cosmos as being linked with a macrocosm to microcosm structure.

This was common to the alchemical text of the hellenistic period and beyond where something ascends and descends the scale of passing through various elements and appeared to be perfected.

It was so difficult to know which text was authentic. Yet the text has been interpreted historically. It may well be that the text appeared in Ismaili circles because they read or wrote the text being described as the being go Hermes Tres Maguestus. Hermes was a type of quasi-angelic prophet of natural science being linked in some way with Idris or enoch who had been ascended to heaven and been transformed into an angle as early as the hekhalet literature. This explains why at the end of many versions of the text Hermes Tres Maguestus claims that his greatness flows from the process so in the first interpretation the text might be describing the mystical transformation of no one less than Hermes.

The alchemical interpretation was followed quickly after with the tablet describing the creation of reality and the perfection of the stone of the philosophers has the ability to perfect nature as the ability to transform base metals into gold. The tablet was an extremely succinct description of the production of the stone of the philosophers which was the lapis philosophorum. This had to be the dominant read of the text through the middle ages starting with early commentaries such as those that accompanied the vulgate from the arabic. When it was first translated into latin and then most early European translations by people known as the hortolanus.

His commentary was important and followed all major intellectuals interested in natural science or alchemy which included folks such as Roger Bacon and Albertus Magnus and then appeared in the aurora conserving one of the most important texts just prior to the fifteenth century.

In Europe, the emerald tablet exploded in the vulgate edition in popularity in the late fifteenth and sixteenth centuries for different reasons.

The first may have been the rediscovery and publication of the corpus hermetic by Marcello related to Hermes Tres

Maguestus or even in Egypt. The printing of the text for the first time with the commentary of ortolanos and a popular compilation de alchemia of 1541 made the text available and inexpensive to obtain. There was a third factor which was that the emerald tablet was increasingly being set into a mythological framework where the text became a type of hermetic version of genesis. This forges a bridge between the patriarchs of the Bible and the later wisdom of the greeks with the emerald tablet that served as the tie that bound them together.

The emerald tablet became a source of primordial wisdom that survived even the flooding and bridging Athens in Jerusalem was the text that became more and more thought of as a compressed source of all wisdom whether it was cosmological, alchemical, metaphysical etc.... We can find all things in the emerald tablet. A third interpretation of the emerald table emerges from Johannes Trithemius and his fellow hermetic philosophers such such as John D and Cornelius Agrippa who understood the text being more something of a metaphysical speculation on nature of reality rather than as a set of laboratory instructions for producing the philosopher's stone. A lot of this interpretation can be found in John D's 1564 Monas hieroglyphica which has to be the most enigmatic texts ever produced and then later in both Rosicrucian and paracelsion literature. It was featured first and foremost in the popular and avant-garde alchemical text Atalanta fugiens though there were dark clouds forming for all of the hermetic by the seventeenth century. It was in the year 1614 that Isaac Cowzban had argued that the corpus hermetic was not as old as once thought and in the 1650s

Athanasius Kicker carried our a similar attack on the emerald tablet arguing the text of the tablet was just a few centuries old, and that is was a fraud.

Despite the terminal condition for the antiquity of the emerald tablet, Isaac Newton composed a translation of the text which survived in his papers he was working on from the vulgate version found in the 1541 day alchemia printed in Nuremberg. It was not until the beginning of the late sixteenth century that the text was associated with a special emblem of paracelsian origin and which served as a type of interpretation of the text but also for deeper alchemical speculation. Despite the attack by the historicist camp, the emerald tablet remains a mainstatin all schools of hermetic or occult thought. The text remained deeply influential through the occult revival containing mystical and occult wisdom. Blavatsky quotes the text as being authoritative in her monumental theosophical text. It was unveiled and perennialists sich as Titus Brookhart has seen the emerald tablet as part of a perennial philosophy. The text was kind of cribbed into one of the manifestos of the surrealist Andrei Breton and received enormous attention from the psychoanalyst Carl Jung.

The emerald tablet has came into full circle from its possible Israeli origins to The New Age Movement where they read the text as maybe a testament to the metaphysical transformation of Hermes Tres Maguestus.

There was a more new age interpretation that individualizes the experience and observes the text as a type of alchemy where one can be transformed into a higher being from cryptic mystical origins to mystic

individualist consumerism. The emerald tablet of Hermes Tres Maguestus seems to be considered as evergreen in interpretation, innovations and has been always somewhere between concealing and revealing. Despite there being enduring interest in the emerald tablet, there has not much in the contemporary literature on the text and when it does it becomes a mess. It received little treatment and important texts such as those of Von Blatto were safe and expect. This means that the only substantial study of the origins of the emerald tablet remained as a Ruska text of 1926 which has never been translated from German. We find it very difficult as well to find the same. There was a recent text in French by Khan who provided an updated account of the reception and commentary tradition of the text through the middle ages but again it was only in French. There has been no major academic study of the emerald tablet in English for about a century. Yet we can still learn about the teachings of the alchemists which we will no go into.

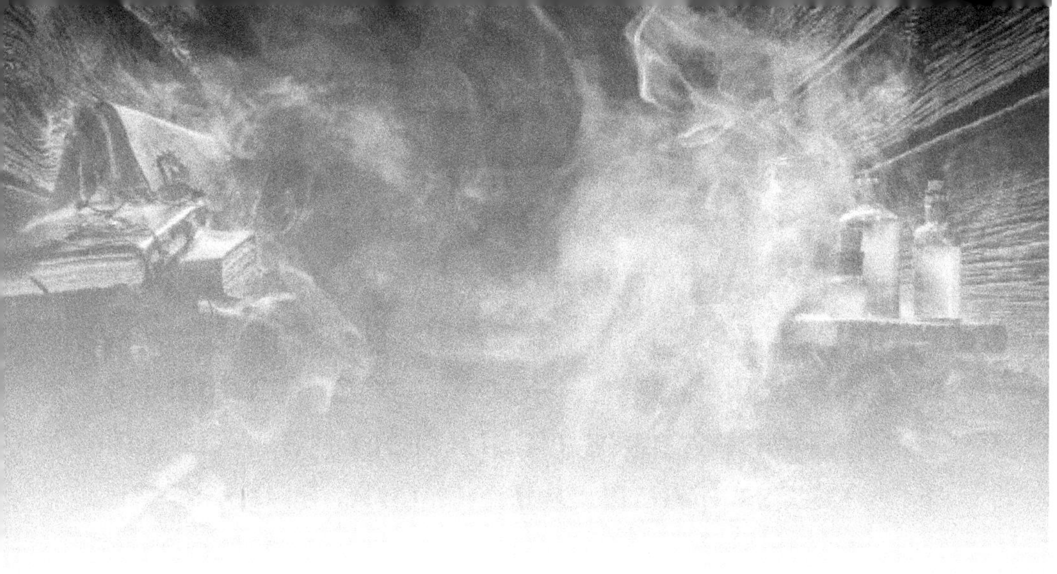

Chapter 13

TEACHINGS TO FATALITIES

We are now going to take a turn of our attention to the lives and individual teachings of the alchemists. The first name found in the history of Alchemy as Hermes Tres Maguestus. He has been regarded as the father of alchemy. Even his name was supplied with a synonym for the Art which we now call The Hermetic Art and even to this today we speak hermetically sealing flasks. The alchemists assumed he was an Egyptian living during ancient times. He is now considered to be purely mythical which is a personification of Thoth who is the Egyptian God of learning, but someone must have written works about him. Of these works, the "Divine Pymander" which is a mysticalreligious treatise was the most important. The "Golden Tractate" which was also attributed to Hermes and which was also an exceedingly obscure alchemistic work was regarded as having been written at a late date.

In the late work to Albertus Magnus, we are told that Alexander the Great has found the tomb of Hermes in a cave near Hebron. The tomb contained an emerald table called "The Smaragdine Table" on which were inscribed the following thirteen sentences based on Phoenician characters:

"1. I speak not fictions things, but what is true and most certain."

"2. What is below is like that which is above, and what is above is like that which is below, to accomplish the miracles of one thing."

"3. And as all things were produced by the medication of one Being, so all things were produced from this one thing by adaptation."

"4. Its father is the Sun, its mother the Moon; the wind carries it in its belly, its nurse is the earth."

"5. It is the cause of all perfection throughout the whole world."

"6. Its power is perfect if it be changed into earth."

"7. Separate the earth from the fire, the subtle from the gross, acting prudently and with judgment."

"8. Ascend with the greatest sagacity from the earth to heaven, and then again descend to the earth, and unite together the powers of things superior and things inferior. Thus you will obtain the glory of the whole world, and all obscurity will fly far away from you."

"9. The thing is the fortitude of all fortitude, because it overcomes all subtle things, and penetrates every solid thing."

"10. Thus were all things created."

"11. Thence am I called Hermes Trismegistus, possessing the three parts of the philosophy of the whole world."

"12. Therefore am I called Hermes Trismegistus, possessing the three parts of the philosophy of the whole world."

"13. That which I had to say concerning the operation of the Sun is completed."

The teachings of these doctrines of alchemistic essence or "One Thing" is everywhere present even penetrating solids and even out of all things for the physical world. Of course, these were adapted and modified over time. The terms Sun and Moon may stand for Spirit and Matter and not gold and silver.

There was another alchemist by the name of Zosimus of Panopolis who flourished during the fifth century and was regarded as the latter alchemists as a master of art. He was written many treatises that deal with alchemy, but there are only fragments that remain.

Based on these fragments, they give us an idea learning about man during his time period. They contain descriptions of apparatus, furnaces, studies of minerals, alloys, of glass making, of mineral waters, and other mystical ideas as well as the ideas of the transmutation of metals. Zosimus was known to state "like begets like". But we don't think all the fragments ascribed to him were really his works. There were other early alchemists such as the Africanus, the Syrian, Synesius, Bishop of Ptolemais, and historian Olympiodorus of Thebes.

During the seventh century, the Arabians had conquered Egypt. This was also the time period when alchemy flourished among them. Geber was regarded as the greatest Arabian alchemist. He was supposed to have lived about the ninth century but his life has not been known. There have been

a large number of works that have been ascribed to him. The majority was not known, but the four Latin MSS which has been printed under the titles "Summa Perfectionis Mettalorum, De Investigation Perfectionis Metallorum, De Investigation Veritstis, and De Fornacibus Construendis" were regarded as genuine.

Based on these works, Geber was regarded as the highest chemist. These describe the preparation of many important chemical compounds. Essential chemical operations such as sublimation, distillation, filtration, crystallization or coagulation as alchemists called it as well as important chemical apparatus such as the water-bath improved furnaces. However as time progressed, it was found that the Summa Perfectionis Mettalorum was a forgery of the fourteenth century, and other works forgeries of a later date.

The original Arabic MSS of Geber has been brought to light. The true writings of Geber have been obscure. They do not give no warrant that believing the famous sulphurmercury theory was due to Geber. It was even proven that he was not the expert chemist he supposed to be. The spurious writings show that pseudo-Geber was an Arabian man of wide chemical knowledge and experience.

Another Arabian alchemist was Avicenna and Rhasis, who lived after Geber, and to whom the sulphur-mercury theory may have been due. The teachings of the Arabian alchemists penetrated into the Western world during the thirteenth century flourished some of the most eminent alchemists who lives and teaching we are now going to talk about. Albert Groot or Albert von Bollstadt was born in Lauingen around 1193. He was educated at Padua and his later years he showed himself acquiring the knowledge of his time. He decided to study theology, philosophy, and natural science and has been celebrated as an Aristolelean philosopher. He has entered the Dominican order and taught publicly at Cologne, Paris and other places. This was made provincial of the order. He held the bishopric of Regensburg conferred to him. He then retired after a few years yo a Dominican cloister where spent time learning philosophy and science. He was a learned man during his time period and a man of noble character. Yet still the authenticity of his alchemistic works has been questioned.

The Dominican Thomas Aquinas may have been a pupil of Albertus Magnus where he learned alchemy. The alchemistic work attributed to him was spurious. According to Thompson's

"History of Chemistry" he was the first to use the term "amalgam" to designate an alloy of mercury with another metal.

Roger Bacon was a medieval chemist, and he was born near Ilchester in Somerset around 1214. He has been said to be regarded as the intellectual originator of experimental research. He studied theology and science at Oxford and at Paris, and he also joined the Franciscan order. He was interested in optics.

It appeared he was acquainted with gunpowder which was still not employed in Europe until many years later. He earned the reputation of being in communication with darkness and suffered much persecution. He was still a believer in the powers of the philosopher's stone to transmute large quantities of base metal to gold and to extend the life of an individual. He felt that "Alchimy" was a science teaching

how to transform metal into another metal and by proper medicine as this appeared in philosophers books. He also felt that "Alchimy" was a science teaching how to make and compound certain medicine called the Elixir which when it was cast upon metals or imperfect bodies fully perfect them. He also believed in astrology, but he was entirely opposed to the magical and superstitious notions held during the time period and his tract "De Secrets Operibus Artis et Nature, et de Nullitate Magic" was an endeavor to prove "miracles" that could be brought by natural science.

Roger Bacon was still a supporter of the sulphur-mercury theory, but he does express surprise that any should employ animal and vegetable substances in the attempts to prepare the Stone. He even states that nothing should be mingled with metals which has not been made from them and that was able to perfect them or to change and a new transmutation of them. There was one process that was necessary for preparing the stone which he tells is "continually concoction" in fire which he felt was the method god has given nature.

Another alchemist was Arnold de Villanova or Villeneuve. He studied medicine at Paris and in the thirteenth century he practiced on Barcelona. He wanted to avoid persecution at the hands of the Inquisition so he left Spain. He found safety with Frederick II in Sicily. He was famous as an alchemist and as a physician. Raymond Lully was the son of a noble Spanish family. He was born in Palma in Majorca about 1235. He had an eccentric character. During his youth, he was a man of pleasure, but during his maturity he was a mystic and ascetic. His career was of adventurous character. During his younger days, he became infatuated with a lady of the name of Ambrosia de Castello who tried to dissuade him from his profane passion. She requested Lully to call upon her and in the presence of her husband her breast was eaten away by cancer. He became actuated by to converting christianity the heathen in Africa. He engaged in the services of an arabian where he might learn the language. The man discovered his master's object, attempted to assassinate him, and Lully escaped with his life. His enthusiasm for missionary work did not abate him. Unfortunately, he was stoned to death by the inhabitants of Bugiah in Algeria in 1315.

There has been a large number of alchemistic, theological, and other treatises that have been attributed to Lully. Many of them were spurious. It was difficult to question to decide which ones were genuine. He derived a knowledge of alchemy from Roger Bacon and Arnold de Villanova. It appeared probable that Lully the alchemist was a personage distinct from the Lully whose life was different than we thought or the alchemistic writings attributed to him have been forgeries similar to psuedo-Geber.

Of those alchemical writings, we have something called the "Clavicula". He states that is the key to all his other books on Alchemy in which the books the whole art is declared. According to this work, there was an alleged method for what he called the multiplication of the "noble" metals rather than transmutation was described in clear language. The stone employed was a compound either of silver or gold. According to Lully, the secret of the philosopher's stone was the extraction of the mercury of silver or gold. He states

in his writings that metals cannot be transmuted and even in the minerals unless they can be reduced into their first matter.

In 1546, there was a work published entitled "Margarita Pretiosa" claimed to be a "faithful abridgement" by Janus Lacinus Therapus, the Calabrian" of a MS. written by Peter Bonus during the fourteenth century. During the life of Bonus, who was an inhabitant of Polo, a seaport of Istria, nothing has been known. But still the "Margarita Pretiosa" was an alchemistic work of great importance. The author brings forward a number of arguments against the validity of the art. He proceeds with arguments in favor of alchemy and puts forward answers to former objections. Bonus was lucid. He felt that metals following the views of pseudo-Geber consist of mercury and sulphur. Even mercury is one and the same where different metals contain different sulphur. He believed that there was two kinds of sulphur that are inward and outward. Sulphur was necessary for the development of mercury but for the final product gold, it was necessary that the outward and impure sulphur should be purged off. Bonus stated that each metal differed from the rest and has certain perfection and completeness on its own, but none of them has reached the highest degree of perfection. For the common metals there was a transit and perfect state that was attained through the slow operation of nature or through sudden transformation power of the stone. The imperfect metals form part of the great plan and the design of nature though they are of transformation into gold. It was also found that a large number of tools and utensils could not be provided if there were no copper, iron, tin, or lead and if all metals were either

silver or gold. They felt that nature had furnished the metallic substance in different stages of development from iron being the lowest state to gold being the highest state of metallic perfection. Nature covered the whole face of earth with water has evolved out of an elementary substance a diversity of forms which embraced the whole animal, vegetable, and mineral world. Nature has differentiated the metallic substance into a variety of species and forms. The Art of Alchemy consists of not reducing imperfect metals to their first substance but in carrying forward nature's work by developing the imperfect metals to perfection and removing sulphur.

Nicholas Flamel was born about 1330 probably in Paris. He took up the trade of a scrivener. Over time, he became a very wealthy man and exhibited considerable munificence. The increase in Flamel's wealth has been attributed to the success in the Hermetic Art. A remarkable book came into the young scrivener's possession which he was not able to understand until he had the good fortune to meet someone as an expert who translated mysteries for him. The book revealed occult secretes of alchemy. Nicholas was able to obtain immense quantities of gold. The story was a legendary nature, and it seems that Flamel's riches resulted from his business as a scrivener and also from moneylending. All of the alchemistic works that were attributed to Flamel were of origin. One of these was entitled "A Short Tract, or Philosophical Summary" was found in "The Hermetic Museum". It was a brief work that supports the sulphur-mercury theory.

The book known as "Triumph-Wagen del Antimonii" was a great book. The author described himself as "Basil Valentine"

who was a Benedictine monk. In his "Practica", another alchemistic work he states he entered the wretchedness of this world and fearful consequences and he had to withdraw from the evil world and devote himself to God. He then relates that entered a monastery but found that he had some time on his hands after performing daily work and devotions. He did not want to waste time in idleness so he took up the study of alchemy which he considered the investigation of those natural secrets by which God has shadowed eternal forces and by the last labors were rewarded by the discovery of a stone that was most potent curing diseases. In "The Triumphal Chariot of Antimony", there were accurate descriptions of antimonial preparations and as Basil had written this work during the fifteenth century the preparations were concluded to have been for his own discoveries. He defends the most utmost vigor the medicinal values of antimony and criticizes far from mild the physicians of the day. Based on this work, Basil Valentine was ranked very high as an experimental chemist, but during his early times its work was regarded as doubtful. Yet still the work of "The Triumphal Charoit of Antimony" gives an accurate account of the knowledge of antimony and the pseudoValentine shows himself to be a man of considerable experience based on this subject.

Isaac of Holland and a countryman with the same name, perhaps his son, was said to have the first dutch alchemists. They lived during the fifteenth century, but nothing was known about their lives. Isaac appeared to have been a practical chemist and his works were in great regard by Paracelsus and many other alchemists. He had the belief that all things in the world were of dual nature which was partly good and which was partly bad. God has created good in the upper part of the world that was perfect and uncorruptical in the heaven and those in lower parts such as beasts, fishes, and sensible creatures, herbs or plants, was indued with a double nature which was perfect and not perfect. The perfect nature was called the Quintessence, and the unperfected was known as the Feces pro dreggs or the venomous or combustible oil. God put a secret nature or influence in every creature and to every nature he gave one common influence whether it was physic or other secret works were found by natural workmanship. There were still things that were not known apparent to our senses. He actually provides directions for extracting the Quintessence for which powers were claimed out of sugar and other organic substances. He was the earliest known writer that mentions the famous

sulphur-mercury-salt theory.

Bernard Trevisan was a French count of the fifteenth century wasted sums of money in the search for stone in which the whole of his life and energies were engaged. He seems to have become the dupe of one charlatan after another charlatan, but as a ripe old age he states his labors were rewarded, and he successfully performed the magnum opus. He speaks of the philosopher's stone as a stone compounded as a body and spirit or a volatile and fixed substance and that is nothing in the world can be generated and brought to light without these two substances similar to male and female. He felt the substances were not of one and the same species yet one stone arise and they appear and are said to be two substances yet it was said to be one. Unfortunately, he has added nothing to the knowledge of chemical

science. Sir George Ripley was another alchemist but he was widely known as an alchemistic philosopher of the fifteenth century who entered upon a monastic life when a youth became one of the canons of regular Bridlington. He carried out some travels and returned to England and obtained leave from the Pope to live in solitude. He decided to devote himself to the study of Hermetic Art.

His main work was "The Compound of Alchymie...containing twelve Gates" was written in 1471. Based on this work, we learn there were twelve processes necessary for the achievement of the magnum opus namely Calcination, Solution, Separation, Conjunction, Putrefaction, Congelation, Cibation, Sublimation, Fermentation, Exaltation, Multiplication, and Projection. Theses were similar to twelve gates of a castle where the philosopher should enter. Ripley popularized the works of Raymond Lully in England, but also does not appear to have added to the knowledge of practice chemistry. He had a book called "Bosom Book" that contained an alleged method for preparing the stone.

Thomas Norton was the author of the celebrated "Ordinall of Alchemy" was born before the commencement of the fifteenth century. The "Ordinall" was written in verse and is found in Ashmole's "Theatric Cheicum Britannicum" was found to be anonymous but the author's identity was revealed by a curious device. Samuel Norton was the grandson of Thomas and was also an alchemist. He states that Thomas Norton was a member of the privy chamber of Edward IV. He had his views regarding the generation of metals. He taught that true knowledge of the Art of Alchemy was obtained by word

of mouth from an adept and in the "Ordinall" he gives an account of his own initiation. He mentions he was instructed by his master which was probably Sir George Ripley and learned the secrets of the Art in forty days when he was twenty-eight years old. He does not appear to have reaped the fruits of the knowledge. Twice he mentions he prepared the Elixir and twice it was then stolen from him after ruining himself and his friends by their unsuccessful experiments.

Let us now take a look at the alchemists after Paracelsus. You might be wondering who exactly is this Paracelsus. His name is probably Paracelsus or to give his correct name it is perhaps Philip, Aureole, Theophrast Bombast von Hohenheim was born in Einsiedelin in Switzerland in 1493. He decided to study the alchemistic and medical arts from this father who was a physician. He then continued his studies later on at the University of Basle. He also spent time studying magic and the occult sciences from Trithemius of Spanheim. Unfortunately, he found the theoretical book learning process of the university curriculum not satisfactory and spent time in the mines. This is where he studied the nature of metals. He spent several years traveling to Europe. He finally returned to Basle who was the chair of the Medical Science of his old university. The works of Isaac of Holland inspired him to improve the medical sciences. During his lectures, he denounced the violent terms of the teachings of Galen and Avicenna who were the authorities on medical matters. He spoke German and his tough manner brought him dislike with the rest of the physicians and municipal authorities who sided with the aggrieved apothecaries and physicians who Paracelsus has exposed. He then fled from Basle and carried out his former roving life. He was considered a alchemist of intemperate habits, but he did accomplish a large number of remarkable cures. Paracelsus combined opposite characteristics that it was difficult to criticize him. It was also a problem of difficulty to determine which books were genuine and what his views on points were.

Paracelsus was one of the first people to recognize the desirability of investigating the physical universe. He taught his students the objective of chemistry was not to make gold, but to prepare medicines. He founded the school of Iatro-chemistry or Medical Chemistry. This combination of chemistry and medicine had benefited science. New possibilities of chemical investigations have opened up which were no longer alchemistic.

Paracelsus's theory was an analogy between man, the microcosm, and the world of macrocosm. He believed all the actions that go on in the human body was a chemical nature. He felt that illness was the result of a disproportion in the body between quantities of three great principles which were the sulphur, mercury, and salt theory that constituted all things. He considered that an excess of sulphur was the cause of fever since sulphur was the fiery principle.

The basis of the intro-chemical doctrines which was the healthy human body was a combination of chemical substances. He believed it was the illness of some change in this combination and curable only by chemical medicines that express a certain truth and a great improvement of the ideas of the ancients. In his discussions of his medical doctrines,

Paracelsus teachings were full of exaggeration and highly ridiculous. The extravagance was pronounced in the alchemistic works attributed to him. He had the belief that an artificial creation of minute living creatures that resemble a human being was called “homunculi” which sounded ridiculous. Yet his writings contain true teachings of a mystical nature. His held his belief that the doctrine of the correspondence of man with the universe was a whole and certain to be true.

It was between the pupils of Paracelsus and the older school of medicine that there was a battle royal which raged for time with a full vindication of Paracelsus’s teachings. But still it was held with the acceptance of the fundamental intro-chemical doctrines. It was necessary to distinguish between the chemists and alchemists and to distinguish those who pursued chemical studies with the objective of preparing medicines and those who pursued studies for their own professional development and also from those whose goal was the transmutation of base metals into gold whether it was a selfish motive or a desire to demonstrate on the physical plane how valid the doctrines of mysticism.

In the following century or two, we find the chemist and alchemist to be united in one as the same person. Glauber and Boyle never doubted the possibility of performing the magnum opus. The alchemists of this period were present as a great diversity. Libavius and van Helmont had much chemical knowledge and skill. Jacob Boehme and Thomas Vaughan were those who stood equally as exponents of mystic wisdom. Then there were those who did not enrich the field of chemistry but masters of the Hermetic Art. There were alchemists of the Edward Kelley and "Cagliostro" type whose main object was their own enrichment at a neighbor's expense. Before discussing the lives and teachings of these men, we will now take a look at the "farfamed" Rosicrucian Society.

The exoteric history of the Rosicrucian Society begins with the year 1614. It was published at Cassel in Germany a pamphlet named "The Discovery of the Fraternity of the Meritorious Order of the Rosy Cross, addressed to the Learned in General and the Governors of Europe." The pamphlet contained general reformation of the world that was accomplished through secret confederacy of the wisest men. The pamphlet proceeds to let the readers know an association exists. It was founded over one hundred years ago by C.R.C. who was the grand initiates in the mysteries of alchemy. The book then concludes by inviting men of time to join the Fraternity. The pamphlet was the cause of considerable interest and excitement.

The following year there was a further pamphlet called "The Confession of the Rosicrucian Fraternity, addressed to the Learned in Europe. In 1616, the was another pamphlet titled "The Chymical Nuptials of Christian Rosencreutz." The latter book is a allegorical romance that describes how an old man who was a lifelong student of the alchemistic art was present at the accomplishment of the magnum opus in the year 1459. There was an enormous amount of controversy. The society had deluded them while others maintained its claims. After four years, people did not get to be interested.

Some writers gifted for romance has seen in the Rosicrucian Society a secret confederacy of antiquity and great powers consisting of great initiates of all ages which is in possession of the arch secrets of alchemistic art. The pamphlets were animated by Lutheran ideals. Luther's seal contained both the cross and the rose which gives the name "Rosicrucian". The accepted theory regarding the pamphlets as a hoax perpetrated by Valentine Andrea. Mr. R.A. Vaughan feels this is a hoax. His scheme was a failure, and he did his best to stop by writing several works in criticism of the society and its claims. Mr. A.E. White rejects the theory and suggests that the Rosicrucian Society may have been identical with the "Militia Crucifera Evangelica" which was a secret society founded by Nuremberg by the Lutheran alchemist and mystic Simon Studion.

Let us now take a look at the lives and teachings of the alchemists. Thomas

Charnock was born at Faversham (Kent) either in the year 1524 or 1526. He did carry out some travels over England where he settled out at Oxford carrying out experiments in alchemy. In the year 1557, he wrote "Breviary of Philosophy". The work is autobiographical that describes Charnock's alchemistic experiences. He mentions he was initiated into the mysteries of the Hermetic Art by James S. Of Salisbury. There was another master who was an old blind man who was on his death-bed decided to instruct Charnock. Thomas was doomed to fail his experiments. On the first attempt, his apparatus caught on fire and everything was destroyed. The next series of experiments were ruined by the negligence of a servant. His final misfortune did not go as well either.

Andreas Libavius was born in Halle in Germany in 1540. This is where he studied medicine and practiced for a short time as a physician. He accepted the fundamental iatrochemical doctrines, but he still criticized the extravagant views expressed by Paracelsus. He was a believer in the transmutation of metals but his own activities were directed to the preparation of new and better medicines. He contributed to the field of chemistry by many discoveries including the preparation of tin tetrachloride. It is still known by the name “spiritus humans Libavii”. Libavius had keen observations and his work on chemistry of the science of his time was regarded as the first textbook of chemistry. It was regarded as high esteem during this time period.

Edward Kelley was born at Worcester in 1555. His life is obscured by traditions that was difficult to arrive at the truth concerning it. The best account will be found in Miss Charlotte Fell Smith’s

"John Dee" in 1909. Edward Kelley was brought up being an apothecary.

He entered Oxford University under a pseudonym of Talbot. He then prated as a notary later on in London. He was said to have committed forgery. He had his ears cropped. To avoid this penalty, he made his escape to Wales. Other crimes he has been accused of include coining and necromancy. He was probably not guilty of all of these crimes, but he was a charlatan. During the time to his escape to Wales, while in the neighborhood of Glastonbury Abbey, he became possessed by a manuscript by St. Dunstan that set forth the grand secrets of alchemy including some of the two transmuting tinctures both white and red.

He was friends with John Dee who was a mathematician, astrology, and interested in experiments in "crystal-gazing". He employed a speculum of polished cannel-coal where he had communication with the inhabitants of spiritual spheres. Kelley did possess some mediumistic powers which were the results of which he augmented by fraud found himself interested in these experiments. He became what is known as the doctor's "scryer", but they also gulled him into believing he was in the possession of the arch-secrets of alchemy. In the year 1583, Kelley along with his learned dupe decided to leave England with their wives and a Polish nobleman who stayed first at Cracovia and then at Prague where Emperor Rudolph II knighted Kelley. There were instances of the belief which the doctor had in Kelley's powers as an alchemist. In his Private Diary in 1586, Dee records that Kelley performed a transmutation for Edward Garland and his brother Francis. He was not always without doubts to Kelley's honesty which has been evident from other entries in his diary. In 1587, there was an event that was recorded to the partner's lasting shame. Kelley informed the doctor that by orders of the spirit which had appeared to him in the crystal, they were to share two wives in common. Kelley's violent temper had been the cause of disagreement between him and the doctor and the incident lead to a further quarrel. In 1589, Emperor Rudolph sent Kelley to prison. It was the price of freedom which was the transmutative secret or a substantial quantity of gold. He was released in 1593, but he passed away in 1595 due to an accident occurred while attempting to escape a second imprisonment.

It was during his incarceration that he decided to write the alchemistic work titled "The Stone of the Philosophers" which consists of quotations from older alchemistic writings. He had earlier works on alchemy that were written.

Henry Khunrath was born in Saxony during the second half of the sixteenth century. He followed Paracelsus's beliefs and traveled to Germany practicing as a physician. Mr. A.E. White describes Khunrath's work "Amphiteathrum Sapientice Aerternce" as purely magical and mystical. Both the date and birthplace of Alexander Sethon who was a Scottish alchemist has not been recorded. Michael Sendivgious was born in Moravia around 1566. Sethon held possessions of the arch-secrets of alchemy. He decided to visit Holland in the year 1602, proceed after a time to Italy, and passed through Basle to Germany. It was said to have performed many transmutations. When he arrived at Dresden, he fell into the clutches of the young Elector Christian II who in order to extort his secret casted him into prison and then put him to the torture. Sendivogius was on a quest to The Philospoher's Stone and stayed at

Dresden, and the hearing of Sethon's imprisonment obtained permission to visit him. Sendivogious offered to help Sethon's escape in return for assisting him in his alchemistic pursuits to which the Scottish alchemist agreed. After there was some money in bribery, Sendivogius's plan to escape was successful. Sethon was freed, but he refused to betray the high secrets of hermetic philosophy to the rescuer. Before he died, he presented him with an ounce of transmutative powder. Sendivogious used up his powder in effecting transmutations and cures and decided to marry Sethon's widow hoping she was in possession of the transmutative secret. He was disappointed that she knew nothing of the matter, but she held the manuscript of an alchemistic work written by her late husband. Later on, Sendivogius printed at Prague a book titled "The New Chemical Light" under the name

"Cosmopolita" which was said to be the work of Sethon. Sendivogious claimed it as his own by inserting his name on the title page in the form of an anagram. The tract "On Sulphur" which was printed at the end of later editions, was said to have been the genuine work of the Moravian. While the powder lasted, Sendivogius traveled and performed many transmutations. He was also twice imprisoned in order to share the secrets of alchemy, on one occasion in which he escaped, and on another occasion obtaining his release from the Emperor Rudolph. After these incidents he appeared to have degenerated into an imposter. But this was probably the case since it was said he was an expert to hide his true character as an alchemistic adept. The book "The New Chemical Light" was held in great esteem by alchemists. The first part treats at length the generation of the metals and The Philosopher's Stone and claims to be based on practical experience. The seed of nature is considered to be one but various products result due to different conditions of development. There is an imaginary conversation between mercury, an alchemist and nature. The second part treats elements and principles.

Michael Maier was born in Rendsberg in Holstein in the year 1568. He studied medicine and became a successful physician. He was ennobled by Rudolf II. He then decided to take up the subject of alchemy and has ruined his health and wasted his fortune in the pursuit of the alchemistic "ignis fatuus" which was the Stone of the Philosophers traveling about Germany and other places to have converse with those who were regarded as experts in the art itself. He also took a part in the famous Rosicrucian controversy where he defended his claims of an alleged society in several tracts. He is noted to be a member of a fraternity and to have himself founded a similar institution. He was a learned man but his work was obscure. He took the time to read an alchemistic meaning into the ancient fables that concern the Egyptian and Greek gods and heroes. Just like other alchemists, he held virtues of mercury in high esteem. In his work "Lusus Serius: or, serious Passetime" he supposes a Parliament of many creatures of the world to meet so that man might choose the noblest of them as king over the rest. The calf, the sheep, the goose,, the oyster, the bee, the silkworm, flax, and mercury are chosen representatives. It was unnecessary that mercy wins the day. His "Subtle Allegory concerning the Secrets of Alchemy" is found in the Hermetic Museum with his "Golden Tripod" consisting of translations of "Valentine's" "Practice" and "Twelve

Keys", Norton's "Ordinal" and Cremer's "Testament".

Jacob Boehme or Behmen was born at Alt Seidenberg which was a village near Gorlitz in 1575. The education he received was rudimentary and when his school days were over, Jacob was apprenticed to a shoemaker. His religious nature caused him to admonish his fellow-apprentices. He traveled as a journeyman shoemaker and returned to Gorlitz in 1594 where he got married and settled in business. He claims to have experienced a wonderful vision in 1598 and then to have a similar vision two years later. The first vision lasted for several days where he believed that he saw the inmost secrets of nature but what appeared dim and vague became clear and coherent in the third vision. He mentions it was vouchsafed to him 1610. He wrote his first book "The Aurora" which he composed himself in order that he should not forget the mysteries that were disclosed to him. During a later period, he produced a large number of treatises of a mysticalreligious nature and has spent his intervening years in improving his early education. The books aroused anger of authorities, and Jacob suffered considerable persecution. He then visited Dresden in 1624 where he had a fever. He returned to Gorlitz and expired in a condition of ecstasy.

We can consider Jacob Boehme as an alchemist of purely transcendental order. He did acquire some knowledge of chemistry during his apprentice days. He employed the language of alchemy with mystical philosophy. Boehme has been regarded as a true mystic, but we think this title is due to Emanuel Swedenborg. Based on Boehme's terminology, The Philosopher's Stone was considered to be "The Spirit of Christ" which should tincture the individual soul.

John Baptist van Helmont was born in Brussels in the year 1577. He devoted himself to the study of medicine at first following Galen and then accepting in part the teachings of Paracelsus. He also helped overthrow old medical doctrines. His chemical researches were great value to science. He was a man who had profound knowledge, a religious temperament, and possess a marked liking of the mystical. He was inspired by the writings of Thomas a Kempis, practiced medicine, and did not ask for a fee for his services. He was a believer in The Philosopher's Stone which claimed to have performed the transmutation of metals on more than one occasion. His theoretical views were fantastical and lived a life devoted to scientific research.

Van Helmont regarded water as a primary element of which all things can be produced. He did not believe that fire was an element or anything material at all. He also did not accept the sulphur-mercury-salt theory. He came up with "gas". Before his time, various gases were looked upon as varieties of air and he made a distinction between gases which could not be condensed and vapors which give liquids on cooling. He investigated a gas what we now know as carbon dioxide (carbonic anhydride) which he termed "gas sylvestre". He did not have suitable apparatuses to collect gases and led to erroneous conclusions.

Francis Mercurius van Helmont was the son of John Baptist. He was born in 1618 and gained the reputation to achieve the magnum opus. He appeared to live luxuriously with a limited income. He was a skilled chemist and physician.

Johann Rudolf Glauber was born in Karlstadt in the year 1604. Little of his life was known. He appears to have traveled about Germany and afterwards visiting Amsterdam. He was patriotic nature and an investigator in the realm of chemistry. He accepted the main intro-chemical doctrines but gave most of his time to applied chemistry. He enriched the field of science with many important discoveries. Crystalized sodium sulphate is still called "Glauber's Salt".

Glauber attributed medicinal powers to the compound. He was also a firm believer in the claims of alchemy.

Thomas Vaughan, who wrote under the name of "Eugenics Philslethes" was born at Newton in Brecknockshire in the year 1622. He went to school at Jesus College in Oxford. He has taken holy orders and to have the living of St.

Bridget's (Brecknockshire) conferred on him. During the civil wars he helped the king but his allegiance to the Royalist lead to him being in trouble. He appeared to have been deprived of his living. He retired to Oxford and decided to study chemical research. He is regarded as an alchemist of the transcendental order. He held his views of the nature of The Philosopher's Stone. He appeared to have carried out experiments in physical alchemy. He passed away in 1666 due to inhaling the fumes of some mercury he was experimenting with.

Thomas Vaughan was a disciple of Cornelius Agrippa who was a sixteenth-century theosophist. He held the peripatetic philosophy in not so good esteem. He was a man devoted to God and had an intense desire for the solution of problems of nature. His works include "Anthroposophia Theomagica",

"Anima Magica Abscondita", "Magia Adamica" or "The Antiquate of Magic". There was a controversy between Vaughan and Henry Moore was marked by acrimony. The use of the pseudonym "Philalethes" was not confined to one alchemist during this time period. The name "Eirenaeus Philalethes" and "Thomas Vaughan" has been confused with each other. The name has also been identified with Dr. Robert Child, but the real identity is still mysterious.

George Starkey or Stirk, who was the son of George Stirk, minister of the Church of England in Bermuda, graduated from Harvard in the year 1646 and then practice medicine in the United States, and then went to England to practice medicine in London. He passed away due to the plague in the year 1665. It was during the years between 1645-1655 that he published "The Marrow of Alchemy" by Eirenaeus Philoponos

Philalethes which some people think he has stolen from his hermetic master. There were other works by “Eirenaeus Philalethes” appeared after Starkey’s death and became popular. Other works include “The Open Entrance to the Closed Palace of the King” and the “Three Treatises” can be found in The Hermetic Museum. There were certain points that he differed from most of the alchemists. He did not agree that fire was an element and that bodies were formed by mixtures of elements. He felt there is one principle in the metals named mercury which arises from the aqueous element and was termed metallically differentiated water. Philalethes’s views as the metallic seed was also of interest. For the seed of gold which he also regarded as the seed of all other metals is something to be separate but may be cut out where metallic seed is diffused throughout the metal and contained in smallest parts. Neither can it be distinguished from its body. It was something difficult to extract.

Besides the teachings of these alchemists, let us look deeper into the history how these teachings came about from the early days to some of the last alchemists. For over two and a half millennia, philosophers and scientists have worked hard to understand what the universe was made of and the principles it operates. Philosophers have engaged in this speculation. The first philosopher to theorize about matters was Thales of Miletus during the sixth century BC which was the Greek city of Asia Minor. According to Thales, there was only one fundamental element, and it was water which was considered to be the material everything was made of. Thales noted the evaporation of water turned into mist and it solidified when it froze. Aristotle felt he had a plausible but incorrect idea.

Thale's had a successor by the name of Anaximander and was said to have been sixty four years old in 546 BC. He agreed that there was one primal material. He did not think it was ever encountered on Earth in its pure state. Anaximander felt that everything in the world should be made of apeiron which was a substance that was infinite and eternal and also a substance that could take on many forms including those of familiar terrestrial materials. The last of the Miletus philosophers was a person by the name of Anaximenes. He created his theory before 494 BC. This was a time period when the Persians destroyed Miletus. Anaximenes did not find Anaximander's ideas to be convincing. He maintained the fundamental element to be air, and air could be condensed into known substances. He felt that there were progressive condensations that condensed it into winds, clouds, water, and into earth and stone.

Empedocles was a philosopher who lived in Agrigentium in southern Sicily during the middle of the fifth centre BC. He knew he performed miracles and that he could control the winds as well as to bring a woman who was dead thirty years back to life. He was also considered to be a leader of a democratic body in his native city. He claimed himself to be a God. Based on legend, he passed away when he jumped into a crater at Mt. Etna to prove that he was a god. It was still uncertain whether he did this or not. Empedocles did not make an attempt to create a new theory of matter. He did his best to reconcile thoughts of those predecessors. He decided to take Thales's theory that everything was made of water and Anaximenes's idea that a primal substance was air. He then added two more elements which was earth and fire. He did not believe that one kind of matter would be transformed into another. Earth could not be changed to water and water could not be changed to Earth. He felt there had to be more than one element. He did not speak of earth, air, fire, and water as elements but he felt they had roots of everything. He felt each was eternal, and they can be mixed in varying proportions to make substances encountered in the terrestrial world. He felt that the elements were combined by love and then separated by strife. His theory was not as mystical as it sounds. Empedocles felt love and strife as physical forces and could act on the particles of matter.Love was a force of attraction.

Empedocle's theory of the four elements dominated Western beliefs for nearly two and a half millennia. It was not until later in the 18th century that his thoughts were disregarded. Aristotle's thoughts and authority were great and often impeded scientific progress. Aristotle added a fifth element of the heavenly bodies. He did agree with Empedocles that earthly objects were made of earth, air, fire, and water. He elaborated on the theory and assigned qualities to the four elements. He felt that fire was hot and dry and air was hot and moist. He also felt that water was cold and moist and earth was cold and dry. It was possible for one element to be transformed into another element. His theories were supported by common observations. He felt that the coldness in water could be made hot and the water would be transformed into air. This appeared to be what happens when water is boiled. When wood was

produced, smoke (air), pitch (water), ash (earth), and fire can be produced. Taking two pieces of flint being struck together a spark can be produced and could be used to kindle a fire. He felt that the fire element was present in rock.

Alchemy was considered to be a fusion of Greek philosophy and the Egyptian chemical arts in Alexandria. Alexandria was the city founded by Alexander the Great near the mouth of the Nile in 331 BC. Egyptians during this period and for centuries practiced embalming, dyeing, glassmaking, and metallurgy. Each of these required a knowledge of some chemical process. There were procedures for making artificial gems and false gold. They thought they could follow Greek philosophy and encounter the Egyptian tradition of practical chemistry. The creation of the field of alchemy showed a step backward.

The Egyptians had known seven metallic elements such as gold, silver, copper, tin, iron, lead, and mercury. They associated these with the seven planets such as The Sun, Moon, Mercury, Venus, Mars, Jupiter, and Saturn. The Greeks did not recognize them as different elements. Based on Aristotle's theory, the metals were mixtures of four elements. He felt the theory that one metal can be transformed into another metal. He just needed to know the chemical procedures that would remove one element and add some more of another or thought would change one element into another element.

The place Alexandria had many different religions and cultures encountered and also a place where different philosophies flourished.

The city was home to Greeks, Egyptians, Jews, and other people who migrated from many places in the Middle East. There were a variety of people such the Zoroastrians, Neoplatonists, Mithraists, Christian Gnostics, and other adherents of other philosophies and faiths. The place Alexandria had believers who really did not believe in Greek rationalism and also wizards and sorcerers, mystics, astrologers, and prophets. By the year 300 AD, Alexandrian alchemy had become mystical because the alchemists have been influenced by current mystical thoughts. They probably thought along these lines since they failed to transform base metals into gold. It was easier for them to dwell on the idea of the spiritual gold of one's soul and then follow long complex procedures to actually make the metal. It was once believed that the Roman Emperor Diocletian in 292 AD that all of the alchemical books should be burned and alchemists should be expelled in Egypt. The story may have been apocryphal. During this time period, alchemy was not known in the Roman west. No decrees were needed at this time.

After Constantine proclaimed christianity to be the cult of the Roman Empire about 330 AD, they felt they should eradicate pagan philosophies including alchemy. They would have succeeded if the Nestorians had not preserved alchemical writings. Nestorius was the leader of the sect and was excommunicated around 430 AD. He then fled to Syria with followers. The Nestorians took a lot of pagan manuscripts and books with them and kept them in the monasteries they founded. Around the year 500 AD, the Nestorians have been expelled from Syria. They then moved to Persia where they founded schools and translated Hellenistic writings into Syrian. One of their subjects at their school was alchemy.

Things were different for the arabic alchemists. The years 640 to 720 were the periods of Muslim conquests. During the end of this period, the Islamic empire had stretched from Spain to Egypt and also from North Africa to Persia. They engaged in wars of expansion. They were not interested to convert the people who they conquered. Remember it was the christians who wanted to eradicate pagan philosophy. Muslims respected learning. Muslim rulers patronized scholars. They even had Greek and Syrian texts translated to Arabic. Arab scholars took time to learn the works of Plato, Aristotle, and other philosophers and alchemy. The Muslims gave alchemy the name. The word is derived from Arabic “alchymia”. The two letters “al” is a Arabic definite article. The origin of "chymia" is still not certain. It used to be thought to be derived from "Khem" which is the ancient name of Egypt. However, other people disagree with this. The Arabs were not interested in mysticism that alchemy has acquired. They pursued it to a more down-toearth manner. This was similar to the early Alexandrian alchemists had done. Centuries later, alchemy reached Europe as a collection of procedures and techniques. The two concepts that were still essential during this time period was "The Philosopher's Stone and "The Elixir of Life". The Philosopher's Stone was a substance that transformed base metals into gold. It was not thought of as a stone but as a "red earth". The Elixir of Life was thought of something that could restore youth and prolong life. It was thought to be made from alchemical gold. There were still practice procedures to make dyes and medicines. There was always more to alchemy than to just make gold. Arabic alchemy was not known in the west until the eleventh century. This was when the translations from Arabic into Latin have been made. Jabir ibn Hayyan who was known as Geber and Abu Bar bin Zakariyya al-Razi was known as Rhazes. More than 2,000 pieces of writings have been attributed to Jabir. Most of them were complied by a Muslim religious sect called "The Faithful Brethren" or "Brethren of Purity". The works were written in different styles. The compilation was completed by the year 1000. The work has been translated under the title "Summa Perfectionis". It was based on the translations of Jabir's writings. Jabir came up with the theory that metals were mixtures of sulfur, mercury, and arsenic except for gold which was made of sulfur and mercury alone. Both the sulfur and mercury Jabir spoke were not substances that were found in nature. They were purified essences that the European alchemists called “philosophical sulfur” and “philosophical mercury”. These were supposed to be not like the common substances. Philosophical sulfur did not burn and gold contained most of the mercury and the least amount of sulfur. Other metals could be transformed to gold if there were ways to increase the mercury content.

Al-Razi’s life was different, and he was also a physician and alchemist in Persia. He wrote a text on alchemy called “Secret of Secrets”. The book was not esoteric in nature. It was a comprehensive and practice laboratory manual that became a tool for European alchemists. The book also contains huge lists of chemicals and minerals and mentions their origin. It describes alchemical apparatus which includes several types of glassware and different chemical techniques.

He did not have an interest for the transmutation of metals as the main goal of alchemy. As a physician, he emphasized medicine and knowing the chemical substances in medicine. The wealth of laboratory techniques were proven to be useful to European alchemists.

It was the appearance of arabic alchemical works in Latin translation that helped European alchemy during the eleventh and twelfth centuries. The European alchemists did not succeed to make gold or the elixir of life, but they did make some important discoveries. It was during the fourteenth century there was an alchemist by the name of False Geber who discovered how to make strong sulfuric and nitric acids. The ancient people and the Arabs only knew about the weak acids such as acetic acid from vinegar and lactic acid from soured milk. They found that strong acids were corrosive and capable of dissolving metals. There was another great work called "The Gloria Mundi", and it implied that anyone who found The Philosopher's Stone would fail to recognize it. There were still alchemists who continued to seek it. They went over cryptic alchemical procedures and performed experiments for the quest of the stone they called "The Great Work".

Alchemical literature was so cryptic. Mercury was not referred to it by its name. It was either called doorkeeper, our balm, our honey, oil, May-dew, mother egg, green lion, bird of Hermes, or other names. Birds flying to heaven were referred to as distillation and a devouring lion meant a strong acid. Marriage might represent certain alchemical procedures. Serpents or dragons may have symbolized matter in the imperfect state. There were

many reasons for writing alchemical procedures. The Church frowned on the practice of alchemy so practitioners may have wanted to maintain some secrecy and also help avoid other dangers too.

For some alchemists, the search for the philosopher's stone was a lifelong quest. Bernard of Treves sought the stone since we was 14 years old and until his death around 85 years old. He wasted so much money during his life time. He was born into a wealth family in either Treves or Padua in the year 1406. He heard stories from his grandfather about the alchemists quest. He became fascinated with the idea seeking the philosopher's stone and studied the works of the Arabian alchemists. He got a hold of his first book "Secret of Secrets". He set up an alchemical laboratory and spent four years and 800 crowns trying to make gold. He was not successful, and then he turned to the works of Jabir. Other alchemists offered to help him. Neither their lore or writings of Jabir brought any success. After about two years, Bernard had spent 2,000 crowns and there was still no success. When Bernard turned 20 years old, he met a Franciscan friar who told stories about Pope John XXII who practiced alchemy and amassed a fortune of 18 million florins while issuing bulls against the competition from other alchemists. Bernard and the friar studied two other alchemists, and they were Johannes de Rupecissa and Johannes de Sacrobosco. They decided to prepare highly distilled "spirit of wine" or alcohol and they thought this might help them achieve transmutation. They distilled alcohol about thirty times, but they only encountered failure. The philosopher stone created did not do anything. Bernard applied alchemical procedures to different materials.

Bernard then came across a magistrate in the city of Treves who believed the philosopher's stone could be obtained from sea salt. He set up an alchemical laboratory on the coast of the Baltic. He worked for a year and a half and worked with the salt, but he only encountered failure even though he repeated the processes five or ten times. When Bernard became forty-eight years old, he still realized he had been seeking the secret of transmutation for more than three decades, but he still did not give up. He decided to travel to Italy, Germany, France, and Spain and seek other alchemists. He encountered a monk by the name of Gottfried Lepor who told him that eggs were the ingredient. Bernard bought 2000 hen's eggs which he and Lepor decided to boil. They shelled the eggs and then heated the shells in a gentle flame until they were white. They separated the whites from the yolks and then putrified them separately in horse droppings. They distilled the material 30 times and obtained a white liquid and a red oil, but it still did not transmute lead into gold.

When Bernard was at Berghem in Flanders, he told the philosopher's stone could be obtained from vinegar and copperas which was green iron sulfate. Bernard experimented with those materials as well. He heard that Master Henry who was the confessor to the Holy Roman Emperor Frederic III achieved success so he headed out to Vienna. Master Henry stated he he had not found the philosopher's stone but claimed he found a method for increasing a quantity of gold. The alchemists who were present should contribute forty-two gold marks. Then in five days, Master Henry said these would increase fivefold. Master Henry began making paste of silver, mercury, and olive oil and placed it in a glass vessel. He decided to hold it over fire and added the forty-two marks before sealing the vessel and then burying it in hot ashes around which the fire was kept up for fifteen or twenty-one days. After that the vessel broke and was found to contain sixteen of the fortytwo gold marks. The other twenty-six had disappeared. Something causes the process to work in the reverse process and Bernard got back four of the ten marks he contributed, while the other alchemists had to share twelve.

At this point in Bernard's life, he was fifty-eight years old. He vowed to give up his alchemical quest. He kept this vow for two months and then he resumed his travels. He decided to travel to Rome and then he went on to Messi a, Cyprus, Greece, and Constantinople. He then traveled to Egypt, Palestine, Persia, and England. The journeys costs him about ten thousand crowns. All of his money was exhausted and then he returned to Treves. Bernard's relatives in Treves considered him to be angry and would not have anything to do with him. He then decided to retire to the island of Rhodes. Bernard met a monk who was also interested learning the secret of transmutation. Both of them did not have the funds to buy the materials to carry out alchemical experiments. He continued to work hard and even lived, slept, and ate in his laboratory. Even though he had exhausted a lot of his money, he continued his quest by reading and rereading alchemical works. Bernard lived in Rhodes until he passed away in 1490. He was still trying to make gold. Yet he did discover the secret of transmutation at a late age and enjoyed his wealth. In his writings, he mentions of not attaining success, and only warns seekers after alchemical truth that they should not be deceived by imposters out there. His last words were

> ***"To make gold, one must start with gold."***

Master Henry conned Bernard and other alchemists out of twenty-six gold crowns. He was one of many frauds out there. Other alchemists had other fraudulent ways to demonstrate they could make gold. One of the favorites was to use a knife or a nail made up of two halves that had to be soldered together. The gold half was covered with varnish and was soluble in alcohol. When the object was dipped into the alcohol solution, the varnish dissolved and made it appear that gold had been created. There were other pseudo alchemists who used double-bottomed crucibles where gold filings had been concealed or dropped pieces of charcoal in which the gold leaf had to be hidden into a crucible. There was still other common deceptions which consisted of using an alloy of gold and mercury so when the substance was heated, the mercury was driven off and gold would be left behind. The frauds then demanded large sums of money in return for the secrets or told the patrons large expenses had to be met if they were to produce gold in significant quantities.

Another story about fraudulent chemists in general during the sixteenth century was an Arab who came to Prague. He got to know the many alchemists where he invited them to a banquet to demonstrate a method for multiplying gold. Everyone who contributed a hundred marks would receive a thousand when the procedure was completed. After he collected the gold from the guests, the host took them to the laboratory where he placed the coins in a crucible along with various alchemical preparations. He then placed the crucible on a fire and then seized a bellows with the intention to make fire burn hotter. All of a sudden there was an explosion which filled the laboratory with coals, smoke, and noxious fumes. The laboratory was still dark. Some of the guests found candles and went back to the laboratory to see if the host was okay. They only found the broken alchemical apparatus and the window opened. The Arab alchemist was gone. Even the two thousand and four hundred marks disappeared with him. It looks like a cheated the guests after all.

Some pseudo alchemists succeeded making large sums of money, whereas others had to suffer less fortunate fates. In the year 1575, there was a woman by the name of Marie Ziegler who was roasted alive in an iron chair. She failed to provide Duke Julius of Brunswick for the procedure of transmutation. In the year 1597, Georg Honnauer promised to transmute iron into gold for the Prince of Wittenberg. He was caught putting gold into his crucibles. Honnauer was hanged on iron gallows. Frederick of Wurtzburg maintained a glided gallows which was reserved for hanging alchemists who failed to keep their promises to make gold. In the year 1402, England passed an act of parliament. It forbade making gold or silver by methods of alchemy. The idea was not to outlaw the practice but give Henry IV who was entitled the right to make gold for people. He hoped that alchemical gold may help him pay state debts. But in 1455, Sir Edmund Trafford and Thomas Ashton were granted the right to make gold. Coins were minted from the product they produced. But the alchemical gold was proved to be an alloy of mercury, copper, and gold.

There were still other pseudoalchemists out there. Around 1390, Chaucer satirized the alchemists "The Canon Yeoman's Tale" just like the English Renaissance poet John Lyly in his comedy "Gallanthea" and Samuel Butler who was an English poet in the seventeenth century in "Hudibras". One of the best-known satires was "The Alchemist". It was a comedy by Shakespeare's rival Ben Johnson who does not target the pseudo alchemists but rather the rich. The play centers on the activities of a butler who poses as an alchemist when the master is absent. The butler's name was Subtle and he had two accomplices which were Face and Doll Common. Subtle swindled people by engaging in quackery and claims to transmute gold. During the conclusion of the play, the master returns unexpectedly and his fraud was exposed. In fact, the character of Subtle could be based on Simon Forman who was mentioned by name of Johnson's plays. Forman was born in 1552 and seems to have been a medical quack who solve love philters as a sideline. He was fined many times for pretending to cure the sick, and he was also sent to prison many times. In 1594, he told fortunes and experimented with transmutation. He attracted wealthy customers who were mostly female. He was also asked to provide philters to the countess of Essex. She wanted to divorce her husband and win the love of the earl of Somerset. The facts came out during a murder trial of a woman who acted as a gobetween for the countess.

There were still alchemical frauds that continued long after alchemy had fallen into disrepute. In the year 1867, there were three frauds who bilked Emperor Franz Joseph for $10,000. In the year 1929, there was a plumber by the name of Franz Tausend who swindled German financiers after convincing them they could make gold from lead. When it was time for Tausend to be arrested, he claimed the method to be based on modern scientific ideas and asked to demonstrate his methods. He was taken to State Mint where he was in the presence of police detectives, the state's attorney, and a judge, he produced a tenth of a gram of gold from one and twothirds grams of lead. Tausend and all of his chemicals and apparatus had been searched for before the demonstration. It appeared the transmutation was real, but the following day it was discovered that the gold was smuggled in a cigarette while he was in prison.

In the year 1701, there was a 19-year-old German apothecary apprentice by the name of Frederick Bottger. He was in need of money to continue his alchemical experiments. He performed fake transmutations in front of his friends. He even stated that if they gave him money to continue his quest, he told them he would repay them more money. Bottger was doing his best to defraud them. He felt he was on the verge of discovering the secret of the philosopher's stone. He was also aware of performing such demonstrations. More than one prince had severe punishments to alchemists who claimed to have produced gold and then failed. The alchemists pledged witnesses of his transmutations to be secret. Rumors were still spreading, but something happened. In October 1701, his employer Frederick Zorn who was the Berlin apothecary released Bottger from his apprenticeship. He was now considered to be a journeyman who worked for wages. During the years of his apprenticeship, Zorn was critical of his alchemical experiments.

There have been alchemists who had been searching for the philosopher's stone without any success for many centuries. Zorn felt that Bottger would do better to master how to prepare medicine than pursue this quest of the philosopher's stone. Bottger also melted down some silver coins and turned them to gold. If Bottger found a way to make gold, he would not have continued his apprenticeship. But this did not occur to Zorn and his friends. They were convinced the transmutation was really true. His friends talk aroused great interest in Bottger.

The Prussian king, Frederick I heard about what happened during the apothecary's shop. Frederick summoned Zorn and questioned him about the transmutation. Frederick seemed to be impressed by Zorn's account.

He ordered the apothecary to come back the next day with his former apprentice. He confiscated the gold that Bottger had made. When Bottger heard of the interview, he realized this was not good news and decided to hide. Bottger failed to appear in the court, so there was a substantial reward for the alchemist's capture. Bother slipped out of the country. He was able to persuade an acquaintance to hide him in a covered wagon and was driven to nearby Saxony.

Bottger decided to enroll as a medical student at the University of Wittenberg. Federick discovered where he was and sent troops to capture this fugitive. Federick could not take back Bottger back to Prussia without the permission of the Saxon authorities. This would have damaged the relations with Augustus who was the elector of Saxony. The Wittenberg authorities were not anxious to let go of Bottger. Tales of his gold making spreader, and it did not do much good for the Prussians to insist the fugitive was a criminal. The Wittenberg authorities sent to the elector and asked for instructions handling the affair. They did not answer immediately because Augustus was also the king of Poland and he was in Warsaw. Many weeks have passed by. The Prussians demanded that Bottger be given custody and Saxon officials continued to delay. A message from Augustus has arrived. Bottger has been ordered to go to prison in Dresden until he revealed the method of making gold. The Saxons knew the Prussian soldiers may become desperate and use force to seize the prisoner while he was en route from Wittenberg to Dresden. They provided Bottger with a military escort.

When Bottger was in Dresden, he was confined in a section of the royal castle equipped with a laboratory. He had three assistants to help him pursue the quest for gold. There were also two members of Augustus's county to supervise the work. Augustus was not patient to witness a transmutation. He ordered the prisoner to send a sample of whatever this philosopher's stone to Warsaw. Bottger could not admit he did not know how to make gold. Even if he did, there was that chance that he could have been tortured. He decided to send a box containing some alchemical apparatus and also some ingredients to Augustus along with instructions for making a small quantity of gold. Bottger's instructions were followed in an experiment that was performed in Augustus's Warsaw palace. All what was produced was a metallic mass that did not even look like gold. But still this did not discouraged Augustus who wanted the alchemist to be confined to carry out further experiments.

Augustus also gave an order for Bottger to be in a more comfortable imprisonment. Bottger responded by giving great promises to the king. He claimed he would produce large quantities of golf about every month. He regretted these promises and was feared of Augustus consequences if he did not make any gold so eventually he escaped from the palace and made his way to meeting a friend with a horse. He rode into Austria and headed towards Prague. His freedom did not last long. Augustus's soldiers traced him to an inn in the town of Inns where he stopped to take some rest. They took him into custody and brought him back to Dresden. The king was still greedy. He believed that Bottger would find a way to produce gold. After consulting members of his court, he spared the young man of a harsh punishment but kept a closer eye on him. In 1705, when Bottger had been a prisoner for more than three years, Augustus demanded his prisoner a date that gold could be made. Bottger still made great claims. He wrote a document to promise to produce gold within sixteen weeks and manufacture two tons of a precious metal during the next eight days. When Bottger failed to keep his promises, the king was furious and had him executed. His advisors casted doubt on Augustus's judgement. He spent large sums of money over a period of years who financed Bottger's experiments. Ehrenfried Walter von Tschirnhaus was one of the advisors. He was also employed to find new mineral deposits and make new manufacturing projects. One of Tschirnhaus's pet projects was to find a way to make porcelain. Bottger continued the project when Tschirnhaus grew old to continue the quest by himself. Bottger was a brilliant chemist and a young man and Augustus listened to this. Even

Augustus was enthusiastic to collect porcelain himself.

Chinese porcelain first appeared in Europe during the sixteenth century. Porcelain was harder than other ceramic material. It exhibited a translucence that no other European pottery could come close to. The first porcelain pieces arrived in Europe and found there way into the way for European rulers. The porcelain trade grew and wealthy aristocrats started to collect objects made of precious materials. Even European potters looked for ways to manufacture porcelain. If they discovered the secrets, the profits could be huge. The secret of making porcelain turned out to be just as elusive as the secret of the philosopher's stone. The translucence of porcelain suggested European potters that the material should be a combination of clay and glass. They tried many different methods of combining glass, clay, and other materials.

Some the methods came out to be superficial. The mistake was that they thought glass was an ingredient. The truth was it was not an ingredient. Chinese porcelain was made by mixing white clay with pulverized stone which contained feldspar and then firing these objects made them from these materials even at high temperatures. During the firing process, the two metals fused together and produced a hard nonporous material.

Tschirnhaus suggested that Bottger work on the porcelain project. The king decided to listen on this. He felt there was no way why Bottger could not find a way to make porcelain while continuing his alchemical experiments. He had Bottger go to the Albrechtsburg which is a royal castle at Meissen. It was nine miles from Dresden and there

was space to set up a larger laboratory. The Albrechtsburg had been pillaged during The Thirty Years War.

Bottger was provided with five assistants and there were twenty-four furnaces built into the laboratory. There were samples of clay from all parts of the kingdom that have been sent to him. Bottger had to keep everything a secret. Bottger did not attempt to make porcelain by mixing clay and glass together. He decided to mix a series of careful experiments where he mixed various different clays with different kinds of rock and fired them even at high temperatures. It was only at high temperatures that the rock be melted. It was a year that Bottger achieved some success. He did not produce porcelain but he did make red stoneware that was finer than what other German potters could make. It was not white, and it also was not translucent. It was a new type of ceramic material, and he expected further progress. His work was suddenly interrupted. Augustus was still king of Poland and was at war with Sweden. He suffered defeat and Swedish troops advanced towards Dresden. Augustus had the most valuable possessions including Bottger transported to Koningstein. When Bottger was there, he still felt like it was a prison and was not allowed to have books, ink, or paper even no matter much Bottger tried to convince the king to have him work on experiments. In 1707, Augustus abdicated as king of Poland and Swedish forces withdrew from Saxony. Augustus ordered that there should be a new laboratory set up in Dresden. When it was finished, Bottger could leave Konigstein. Bottger felt at this time that if he failed to produce with gold or porcelain he would be executed.

Bottger decided it would be better to make porcelain than transmuting base metals into gold. He carried out his experiments with different combinations of clay and minerals. He used China clay which contained feldspar, but Bottger still did not achieve anything because the temperatures of his furnaces were too low to melt the material. He decided to use alabaster which is a type of gypsum that is snow white and translucent. He mixed clay and alabaster in different ratios and that if he used seven to nine parts of clay to one of alabaster, there should be a hard white translucent material. He succeeded making porcelain. But they were still not that high quality made like in China. He still did not know how to produce glaze, but he was still confident in his techniques to produce porcelain of finer quality. Augustus was pleased but still kept Bottger in prison until he could make gold. The alchemist was still in Dresden even though a factory was set up at Meissen. He was also appointed as director. Augustus made Bottger a baron in the year 1711. The alchemist lived a life of an aristocratic gentleman. Augustus gained control of the polish crown in 1710 and still didn't want to let him go until he could find the philosopher's stone. Bottger became very ill, and we learned he was a heavy drinker in the past. The latter part of his life he rarely spent a day in sober. Anyone who labored in an alchemical laboratory inhaled poisonous fumes especially arsenic and mercury. These were used in alchemical experiments during this time period. He was freed later on but even though his health seemed to improve, it was not true. He passed away in 1719. He spent more than twelve years as a prisoner and only spent five years left for freedom.

The never-ending quest continued. Alchemy was superseded by chemistry in the 18th century. Alchemical practices nearly died out. There are still people until this day that practice this art. Even someone can study alchemist at Paracelsus College in Australia. Modern chemists do not attempt to find a way to make gold. The mystical alchemy still exists today. It is called "esoteric alchemy" and becomes intermixed with the "new age" and mystical ideas. The various Rosicrucian groups make use of alchemical concepts and mysticism. There were herbal remedies sometimes made by alchemical methods. Sometimes, they say alchemy as a path to spiritual growth. It is the spiritual gold that is sought. But the days of alchemy began to decline towards the beginning of the 18th century.

Chapter 14

THE LAST ALCHEMIST

"The vital thing is not the transmutation of the metals, but that of the experimenter himself. It is an ancient secret that is rediscovered by a few people every century."

Fulcanelli Fulcanelli was often referred to as one of the last alchemists. He was said to create the philosopher's tone or the elixir of life to achieve immortality. The name Fulcanelli was used by a French alchemist and as an esoteric author whose identity has been debated. We can think of the name Fulcanelli as portions of words. Falcon was the ancient God of Fire which includes the letter I being a Canaanitename for God which also means sacred fire. The appeal of Fulcanelli was a cultural phenomenon which was due to the mystery of his life and also his disappearance.

He was known to have performed a transmutation of lead into gold two times. The first took place in 1922 together with a devoted pupil Eugene when both of them performed a successful transmutation of 100 grams of lead into gold. This was in the

presence of Julian Champagne and Gaston Sauvage. The demonstration took place in a laboratory setting of Gas Works of Georgie company at Sarcells. This was achieved with a small quantity of projection powder which was also known as the philosopher's stone prepared by Fulcanelli. The second place took place in the year 1937 at the Chateau De La Rey. When Fulcanelli performed a transmutation of 225 grams of lead into gold and 100 grams of silver into uranium before different witnesses which includes a chemist, two physicists, and a geologist. After these demonstrations, Fulcanelli disappeared completely and bestowed increasing scientific discoveries with references to secret ancient knowledge for learning the forces of the material world. He learned this based on two hermetic books such as "The Mystery of the Cathedrals" and the dwellings of philosophers published by students after he mysteriously disappeared in 1926.

The work that he was known for attracted many seekers who believed in the power of magic and understood ancient knowledge was preserved by unknown individuals who have managed to master their lives. Fulecaneli was considered to be Frenchman educated in alchemical law, architecture, art, science, and languages. Theories about Fulcaneli speculate he may have been a French occultist. He may have been a member of a former royal family of the house of Valois or another member of brothers of heliopolis which was a society centered around Fulcaneli. There have been many theories that has been made to solve this mystery. He was known to be associated with mythical and timeless counts in German but his identity remains unknown. There was

no one out there who knew who was Fulcaneli, but he was known for his great work and even greater legend.

There have been dedicated seekers who have tried to identify where he went, but they did not have any luck. Even when he explained the alchemical process of transforming metal into gold and then overcoming death, Fulcaneli confused the interlocutor by leading him back from the alchemical process to his own identity. Fulcaneli did his best to explain the great work by emphasizing the purpose of alchemy and completing the great work to create gold but to transform oneself to a higher purified state. Accomplishing this great work represents the culmination of a spiritual path, the attainment of Enlightenment, or rising the human soul from unconscious forces that bind it. The great work signifies the spiritual path towards self-transcendence. This was the process to bring unconscious complexes into conscious awareness to integrate them back to oneself. The transformation was symbolized by alchemical processes of transmuting lead into gold. This can be thought of as representing a base spirit, and the goal was to transmute oneself into gold. It was considered to be the divine spirit that represented the reuniting of the soul with the divine. Alchemists believed the process has enabled them to make gold within themselves. They decided to equate gold with the sun and sought to manifest radiance within their own beings to create the internal sun. Gold was the metal of the sun, and this has been considered to be many as crystallized sunlight.

When gold was mentioned in alchemical tracks, it might be either the metal or the celestial orb which was the source or spirit of gold as gold was

considered to be the symbol of spirit where the base metals represented ones lower nature.

Certain alchemists were also called miners, and they were pictured with different picks and shovels digging into the Earth in order to search for the precious metal. There were some people who used a pearl hidden in the shell of an oyster towards the bottom of the sea to signify spiritual powers. The seeker after truth became what we call the pearl fisher. He descended into a sea of material illusion to search for understanding when the alchemist stated that every animate and inanimate thing in the universe that contained the seeds of gold. What they meant was the grains of sand possessed a spiritual nature for gold and was also considered to be the spirit of all things.

"The purpose of alchemy was not to make something out of nothing but rather to fertilize and nurture the seed which was already present. Its processes did not actually create gold but rather made the ever-present seed of gold grow and flourish."
P. Hall

Just before the first world war, Fulcanelli met another French alchemist during this time period whose name was Schwala de Lubic. There was a day when Fulcanelli approached Schwala at a cafe telling him he heard about him and also invited to stop by his house to discuss topics that he did not feel comfortable talking about in public. When they met, Fulcanelli let Schwala know about a manuscript that he stole from a Paris bookshop. While he was cataloging an ancient book for a bookseller, Fulcanelli came across a peculiar piece of writing a six-page

manuscript which described the color of the alchemical process. Both Fulcanelli and Schwala met often and talked about the great work of transmutation of both spirit and matter. Schwala left Paris since he felt distracted. He moved to South of France where he invited Fulcanelli to join him to talk amore about alchemy. They performed a successful opus which involved the secrets of an alchemical stained glass such as the red and blues of the rose windows of cathedrals. Both Schwala and Fulcanelli may have came up with the formula:

> ***"By pre-arrangement, this was intended to be their last meeting, and all traces of their association were to be eliminated, never to be mentioned again the monthly stipend would cease, and Fulcanelli was to leave directly."***
> **Andre Vanden Broeck, Al-Kemi**

This was supposed to be the last meeting and Fulcanelli went his own way.

> ***"There was to be no discussion, no further conversation after the experiment, and regardless of success or failure, no subsequent meeting was envisioned the driver was waiting, the luggage loaded."***
> **Andre Vandenbroeck, Al-Kemi**

> ***"Schwaller noticed the new spark in Fulcanelli's eyes, the new power and his bearing, they parted as strangers to each other in all ways, including their interpretation of what they had made come to pass."***
> **Andre Vandenbroeck, Al-Kemi**

Unfortunately, there was tension between them, and there was suspicion that Fulcanelli stole the manuscript from a bookseller. The ideas of his famous work of the mystery of the

cathedrals was said to have been taken from Schwala De Lubic. Schwala at this time had been diligently working on documenting the Hermetic symbolism which was found in European Cathedrals. Fulcanelli became aware of Schwala's work about deciphering cathedral symbolism. Schwala intended to publish the book.

Fulcanelli always proudly talked about the connections he had with the publishing world. He expressed interest to borrow Schwala's first draft to assess its suitability for publishing. Schwala trusted Fulcanelli and agreed nothing more when the draft took longer than expected to be returned. When Schwala received the draft back, Fulcanelli insisted the information should not be published as it could reveal too much and could lead to adverse consequences. Schwala had the same thoughts as Fulcanelli about sharing discoveries. Schwala gave no thought about this interaction until many years later than when he found out a book called "The Mystery of the Cathedrals" had been published. The story remains as a myth. Fulcanelli's primary work manifests as this book was written in the year 1922 and was also unveiled to the public in Paris during the year 1926.

Fulcanelli suggested a similarity pointing out that like the Egyptian pyramids there hold a series of mysteries revealing the architecture and engineering of medieval gothic cathedrals which conceals occult knowledge. He believed that the structures had a purpose beyond dedication of religion. He felt they were repositories of divine secrets which housed profound musings of the forefathers pertaining to philosophy, religion, and society cathedrals just like sanctuaries in general possess a welcoming essence offering refuge to those burdened with disgrace.

The sacred spaces served as realms for spiritual enlightenment and an initiation of spiritual seekers. According to Fulcanelli, the term gothic came from art goth which was a particular form of language which allows others to interpret thoughts that would become what we call the language of the Freemasons who took the time to build the cathedrals and transferred the secret understood using the language of christ. The cathedrals become a spoken Kabbalah. According to Fulcanelli, the phonetic Kabbalah was a special use of language which drew phonetic similarities and other symbolic techniques to expand expressions of words. Note that the phonetic Kabbalah was not the Hebrew Kabbalah. The phonetic Kabbalah was derived from Latin cabals, and it was based on phonetic assonance and resonance to echo the science of the words of the ancient Greek nomenclature. This was based on Walter Lang who wrote an introduction of the English translation of Fulcanelli's mystery of the cathedrals. The basic principles of the phonetic Kabbalah were restored in Fulcanelli's work. He states the smallest details in the cathedrals hide based on occult meaning. It was the cabalistic art to conceal this arcane meaning to represent and create a link with the creator everything inside cathedrals was adorned with gold and painted in vibrant colors which symbolized the gates of paradise.

The temple featured three main colors which were white that signified purity and a sought after light for the initiated, black which represents evil or associated with darkness, and red symbolizing the culmination of spiritual journey where spirit prevails over the physical body. Together the three colors were referred to as oriflama. They were also known as the triple colors of work they represent the path from darkness to illumination and then to the spiritual realm.

His secondary work which was on the dwellings of a philosopher was published in 1929. It shared a similarity of essence of his primary work. This was a compelling masterpiece in which Fulcanelli delves into the realm of medieval castles and dwells analyzing the architectural forms architectural proportions that painted windows and carvings. Fulcanelli presents his findings and proves the validity of the facts. He blends historical narratives, an analysis of architecture, and esoteric knowledge that created a masterpiece that was both stimulating and captivating. The two books were important alchemical works for the past two centuries. His masterpiece takes the reader through alchemical labyrinth that decodes the monuments and architectural decoration built by those who have engaged in great work. Fucanelli instructs by showing the history should be interpreted the proper way. He decodes and interprets many different alchemical symbols of the houses of the alchemist and philosophers, but he goes to great lengths to explain the alchemical view of past centuries. Fulcanelli presents those with deep mysteries of the great work.

Jacques Burger and Louis Pauwels wrote the book "The Morning of the Magicians" in which they link alchemy to atomic physics. They suggested the early alchemists knew more than actual function of atoms that was assumed and explored topics such as crypto history, ufology, occultism in Nazism, spiritual philosophy, and dear glocker which was a topsecret scientific technological device developed in the 1940s in Nazi Germany. This book includes a discussion of a significant encounter that occurred in 1953 which was nine years after the disappearance of Fulcanelli where one of the authors Louis Powells took the time to first meet an alchemist.

> ***"It was at the cafe pro cope in Paris, which was then coming into fashion again. A famous poet, during the time I was writing my book on Gurdjieff, had arranged the meeting, and I was often to see this singular man again, though I never succeeded in penetrating his secrets."***
> **Louis Pauwels**

After engaging in a conversation, Louis Powell asked this mystery man if he knew anything about Fulcanelli.

The man states despite Fulcanelli's disappearance, he was not dead

> ***"It is possible to live infinitely longer than an unawakened man could believe. And one's appearance can change completely. I know this; my eyes know it. I also know that there is such a thing as the philosopher's stone."***
> **Louis Pauwels**

Alchemy in their view could be an important relic of a science, a technology, and a philosophy that belonged to civilization which had disappeared. What they discovered in alchemy in light of contemporary knowledge did not lead them to believe the techniques to be subtle and precise which could could have been the result of a divine revelation fallen from heaven and not that they have rejected the notion of revelation but what has been read about the saints and the

great mystics. They have never noticed god speak to those in some technical language.

> ***"Place thy crucible, o my son, tinder polarized light! Rinse out the slag in water thrice distilled!"***

Alchemy was thought of as fragments of a science that has lost fragments in the absence of context that it was difficult to understand or to make use of progress from this point which must be halting but in a definite direction. There was a profusion of moral and religious interpretations. The fragments that have been preserved maintained secrecy. Fulcanelli believed that alchemy was the link with civilizations which disappeared thousands of years ago, and there were archaeologists that knew nothing. There was no archaeologist or historian that will admit civilizations have existed in the past to be more advanced than current science and techniques.

Advanced techniques and scientific knowledge simplified the machinery and traces of what was accomplished. There would be no historian or archaeologist who had not been thorough in scientific education and which could carry out the researches and explorations which would throw any light on these matters. The strict segregation of many disciplines necessitated by fabulous advances in modern science was concealed from other fabulous discoveries of an earlier age.

If there were advanced ancient civilizations in the past, then we may ask where did go and why did they disappear. If you fail to keep the great power in check, then there was a great destruction which would follow. Eugene Cancellier who was a disciple of Fulcanelli and one of the leading

specialists on alchemy was struck by a passage which Jacques Belgier had written as a preface to the bibliotheque Mondial. It was an anthology of the sixteenth century poetry. In the preface, Belgier alluded to alchemists and the cult of secrecy. He wrote the following:

> ***"On this particular point it is difficult not to agree with them. If there is a recipe for producing hydrogen bombs on a kitchen stove, it is clearly preferable that this recipe should not be disclosed."***
> **Jacques Bergier**

There was a meeting with Jax Bergier and Fulcanelli that occurred during June 1937 in a laboratory of the gas board in Paris. Fulcanelli actually communicated with Jax Bergier and warns French nuclear physicist Andre Hellbronner use of nuclear weapons. According to Fulcanelli, nuclear weapons had been used before and also against humanity. Andrei Hellbronner who was a nuclear physicist, a French occultist, and Grand Master among others were assassinated by Gestapo towards the end of World War II.

Jax Bergier claimed eight years before the first atomic bomb test in New Mexico he was approached by some mysterious person and asked to deliver a message to Andre Elbron. Who ever this man was mentioned that he felt it was his duty to warn scientists of the danger of nuclear energy since alchemists of the past had obtained the secret knowledge and destroyed them. Jax Bergier felt the mysterious man to be Fulcanelli. The following is a verbatim translation from an original transcript:

"You're on the brink of success, as indeed are several others of our scientists today. Please, allow me be very very careful. I warn you.. the liberation of nuclear power is easier than you think and the radioactivity artificially produced can poison the atmosphere of our planet in a very short time: A few years. Moreover, atomic explosives can be produced from a few grains of metal powerful enough to destroy whole cities. I'm telling you this for a fact: The alchemy have known it for a very long time..."

"I shall not attempt to prove to you what I'm now going to say but I ask you to repeat it to Mr. Helbronner: certain geometrical arrangements of highly purified materials are enough to release atomic forces without having recourse to either electricity or vacuum techniques..."

"The Secret of Alchemy is this: There is a way of manipulating matter and energy so as to produce what modern scientists call 'a field of force'. The field acts on the observer and puts him in a privileged position face to face with the universe. From this position he has access to the realities which are ordinarily hidden from us by time and space, matter and energy. This is what we call The Great Work".

Fulcanelli mentioned the element plutonium which was a substance not isolated until February 1941. It was officially named unit March 1942. This was not consistent five years after Bergier's encounter which deepened the mystery. The story captured the attention of the American office for strategic services which was the precursor to the CIA. They launched

an investigation to find Fulcanelli after the war, but they failed to locate him. According to Eugene Canceller who was a disciple of Fulcanelli, the last encounter with Fulcanelli happened in 1953 which was years after the disappearance when he visited Spain and taken to a castle high in the mountains. Cancellier had known Fulcanelli as an elderly man in his 80s. He was androgynous which was the final stage of alchemical transformation which Fulcanelli called the divine and regain like in the hermetic creation myth when God creased the cosmos and then God created man and rogynous in his image.

This suggested Fulcanelli achieved a divine transformation which allowed him to revert back to his true form. The reunion itself was brief and Fulcanelli once again disappeared. There were no trace of his whereabouts.

Fulcanelli was popularized by Eugene Cancellier. Yet there were people who believed he was a fraudster while others believed he was the real Fulcanelli in disguise. The true answer remains a mystery, and no one will know for certain, but this legend of Fulcanelli whether the story was true or not there were still valuable truths and knowledge within these tales. It would be noteworthy to mention the alchemical teachings and the cautionary message about the nuclear weapons. For alchemists, it should never be forgotten that power over matter and energy was only a secondary reality. The real aim of alchemist's activities which were the remains of old science belonged to a civilization long extinct was the transformation of the alchemist to a higher state of consciousness. The material resulted as something spiritual. Everything becomes oriented towards the transmutation of one towards the deification of fusion with divine energy, and the fixed center from which all material energies emanate. It was a science that tended to exalt one rather than matter as the following states:

> ***"The real aim of physics should be to integrate man as a totality in a coherent representation of the world."***
> **Telihard De Chargin**

The science of alchemy came a long way, and in time the days of alchemy would come to an end which would lead to the beginnings of a new modern age of chemistry.

Chapter 15

THE END OF ALCHEMY

The alchemists worked hard in their search for the Stones of the Philosophers, and we wonder if they were successful achieving transmutation. Many apparent transmutations may have occurred and observers were either self-deceived by a superficial examination that there were certain alloys that resembled the noble metals or were cheated by imposters. We should still not assume that real transmutations did not take place. Modern research tells us that it may be possible to transmute other metals like lead or bismuth into gold. Out of the many experiments carried out, real transmutations take place. Even if the quantity of gold could have been formed, it would only be formed in the smallest amount during the ancient days of alchemy. If there was one method where metals may be transmuted, then there should be other methods. John Baptist van Helmont was a physician and chemist and due to his nobility of character he carried out the transmutation of mercury into gold. The composition of the stone was still unknown to him. John Frederick Helvetius was an excellent doctor of medicine and a physician. He published at the Hague in the year 1667 the transmutation he effected.

He described the glassy substance as a pale yellow color. He obtained The Philosopher's Stone and performed a transmutation. His gold was assayed, and the gold was further tested.

Even his experiments tells us not to be sure that a real transmutation has taken place. An agnostic position seems to be more philosophical. Even if the alchemists didn't discover the Grand Arcanum of Nature, they discovered other scientifically important facts. The genesis of chemistry speaks for itself. Even if they did not prepare The Philosopher's Stone, they prepared a large number of important chemical compounds. The labors were the seeds for which modern chemistry developed. The highly important science was included in the expression "The Outcome of Alchemy".

It was the intra-chemists who first investigated chemical matters with an objective aim being the preparation of useful medicines. We can think of the medical-chemist and the alchemist as being one person just like Paracelsus. It was later on that chemistry was recognized as a distinct science separate from medicine.

But alchemy was also considered to be a distressing nature. Alchemy was also considered to be a cloak for fraud and those who wanted to become alchemists wanted to accumulate wealth in a short time but it did not always work out. There was evidence that the degeneracy of alchemy commenced as early as the fourteenth century, but still the attainment of the magnum opus was possible for three or more centuries. These alchemistic promises of health, wealth, happiness, and a pseudo-mystical language was employed by these imposters. Some of their tricks such as the use of hollow

stirring rods which gold was concealed convinced the public of their claims. We have already talked about the pseudo-alchemist Edward Kelley. There was another pseudo-alchemist by the name of Count Cagliostro. He was definitely a charlatan. We still do not know who his true name was, but we could identify him with the Italian swindler Giuseppe Balsamo who was born at Palermo in the year 1743 or maybe 1748. He apparently disappeared thirty years later of which the majority was spent committing various crimes. According to the author W.R.H Trowbridge in his book "Caglio: The Splendor and Misery of a Master of Magic", he felt he was an honest and sincere in his beliefs and in his actions. As a philosopher, it was only being discovered how great he was. He claimed to have converse with spiritual beings. However, the results of modern psychical research had robbed claims of impossibility.

The earlier portion of Cagliostro's life was not known. The latter portion was overlaid with legends and lies such that it was impossible to get to the truth that concerns it. It was the year 1776 when Cagliostro and his wife were in London. He became a Freemason joining a lodge that was connected to "The Order of Strict Observance" which was a secret society incorporated with freemasonry and which was concerned with occult subjects. Cagliostro was not satisfied with the rituals and devised a new system that he called Egyptian Masonry. He taught that Egyptian Masonry was to reform the whole world. He set himself out leaving England for the Continent to convert Masons and others based on his views. But we should still look for his motive power of his career without looking for any true knowledge of the occult. It was least probable that he possessed hypnotic and clairvoyant powers. This was the case based on his visit to Russia. At St. Petersburg, he gave a good show and at Warsaw he was detected in the process of introducing gold in a crucible that contains the base metal he was about to transmute. At Strasburg in the year 1780, he ended up being more successful. He appeared to be a miraculous healer of all diseases, though whether his cures were described to some as simple but efficacious medicine which he discovered to hypnotism to the power of the imagination on his patients or to the power of imagination who have recorded the alleged cures.

When he was at Strasburg, Cagliostro came into contact with Cardinal de Rohan where they became friends but in the end proved Cagliostro's ruin. He then visited Bordeaux and Lyons where he founded lodges of Egyptian Masonry. He then visited Paris where he reached the height of his fame. He then became extraordinarily rich and he accepted no fee for his services as a healer. But there was still a substantial entrance-fee to the mysteries of Egyptian Masonry with its alchemistic promises of health and wealth ended up being prospered. At the peak of his career, things went wrong. He was friends with de Rohan and ended up being arrested in the connection with The Diamond Necklace Affair with the famous Countess de Lamonte. Whatever he was guilty of, he was innocent of the charges. After being imprisoned in the Bastille for many months, he was then tried by the French Parliament to be pronounced innocent and released. But the king still banished him and he left Paris for London where he has been persecuted by agents of the French king. He then returned to the continent where he reached Italy. He was then arrested by the Inquisition and then condemned to death on a charge of being a Freemason. This was a dire offense seen by the Roman Catholic Church. The sentence was modified to perpetual imprisonment. He was then confined in the Castle of San Leo where he passed away in 1795 after being imprisoned for four years.

We have came a long way learning about the history of alchemy and mystical sciences to the present. By 1000 BC, civilizations used ancient technologies that would form the basis of various branches of chemistry such as the discovery of fire, extracting metals from ores, making pottery and glazes, fermenting wine and beer, extracting chemicals from plants for perfume and medicine, rendering fat into soap, making glass as well as making alloys such as bronze. Alchemy was the beginning of chemistry even though it was unsuccessful in explaining the nature of matter and its various transformations. By performing experiments and then recording the results, alchemists set the stage for modern chemistry.

We have also learned in our earlier discussion how humans played a role in alchemy. The first chemical reaction used in a controlled manner was considered to be fire. Several people have thought fire was seen as a mystical force that brought one substance into another such as burning wood or boiling water while producing both heat and light. Fire had affected many aspects from the early societies. These included activities such as cooking and habitat heating and lighting to even the most advanced uses such as making pottery and bricks and the melting of metals to make tools. It was actually fire that led to the discovery of glass and the idea of purifying metals which lead to the rise of metallurgy.

Chemistry even goes back to a 100,000-yearold ochre processing workshop that was found at Blombos Cave in South Africa. Speaking about ochre, it was considered to be a natural clay earth pigment consisting of a mixture of ferric oxide and varying amounts of clay and sand. It ranges in different colors from yellow to deep orange or brown. It was also the name of the colors produced by the pigment that is a light brownish-yellow. There was also a variant of ochre that contained a large amount of hematite or dehydrated iron oxide which had a reddish tint known as red ochre. In fact, the word ochre even described clays colored with iron oxide derived during the extraction of tin and copper.

As mentioned, ochre was considered to be a family of earth pigments that included yellow ochre, red ochre, purple ochre, Sienna, and umber. The major ingredient of ochres was iron(III) oxide-hydroxide which was also known as limonite which gives a yellow color. There was also a range of minerals that may also be included in the mixture. Yellow ochre or gold ochre, $FeO(OH) \cdot nH_2O$ is a hydrated ion hydroxide (limonite). Red ochre, $Fe_2O_3 \cdot nH_2O$ has a red color from the mineral hematite which is also an iron oxide that is reddish brown when hydrated. Purple ochre is also a rare variant that is identical to red ochre by chemical means but has a different color caused by different light diffraction properties associated with greater average particle size. Brown ochre, FeO(OH), or goethite is a partly hydrated iron oxide. Likewise, lepidocrocite which is known as gamma-FeO(OH), is a secondary mineral. It is a product of the oxidation of iron ore minerals that is found in brown iron ores.

Sienna has both limonite and a small amount of manganese oxide which is less than 5% and which makes it darker than ochre. Umber pigments have a larger proportion of manganese which is about 5 - 20% which also makes them dark brown. Heating natural Sienna and umber pigments causes them to be dehydrated and some of the limonite gets transformed into hematite which gives them reddish colors called burnt Sienna and burnt umber. Ochres are considered to be non-toxic and can be used to make oil paint that dries quickly and also covers surfaces thoroughly.

The existence of ochre was evidence to show that humans had some elementary knowledge of chemistry. There were paintings drawn by early human beings consisting of early human beings mixing animal blood with other liquids found on different cave walls which showed that there was some small knowledge of chemistry.

There was also the rise of early metallurgy. We have talked about the earliest recorded metal employed by human beings which was gold. Gold can be found free or in its native form. There have been small amounts of natural gold to be found in Spanish caves that were used during the late Paleolithic period around 40,000 BC. Even silver, copper, tin, and meteroric iron has been found to be native allowing a limited amount of metal working during the ancient cultures. Egyptian weapons were of meteoric iron in about 3000 BC and were highly prized as "daggers from Heaven". During the early stages of metallurgy, there were methods of purification of metals, and as mentioned in our earlier discussion that gold in ancient Egypt as early as 2900 BC became a precious metal.

As we approach The Bronze Age, early human beings discovered that certain metals can be recovered from their ores by heating rocks in a fire such as tin, lead, and even at higher temperatures of copper. The process was known as smelting. There was first evidence of this extractive metallurgy that dated from the 6th and 5th millennia BC. They were found in the archaeological sites of the Vinca culture, Majdanpek, Jarmovac and Plocnik in Serbia. There was early copper smelting found at the Beloved site. These examples actually included a copper axe from 5500 BC. There were other signs of early metals found from the third millennium BC at places such as Palmela (Portugal), Los Millares (Spain), and Stonehenge (United Kingdom). It was not easy to define the ultimate beginnings and there were many new discoveries continuing to be made.

It does make sense that during The Bronze Age bronze actually existed. The first metals were single elements or combinations of elements. Combining copper and tin creates an alloy called bronze. This made a major technological shift which began The Bronze Age about 3500 BC. The Bronze Age was a period of time when even the most advanced metalworking included techniques such as smelting of copper and tin from natural occurring outcroppings of copper ores, and then by smelting those ores to cast bronze. A lot of these naturally occurring ores included arsenic as a common impurity. There were copper/tin ores that were rare as reflected in the absence of tin bronzes found in Western Asia before 3000 BC. After The Bronze Age, the invention of the metallurgy was marked by armies seeking strong weaponry. The state in Eurasia prospered when they made superior alloys which made better armor and better weapons. Much progress in metallurgy and alchemy was made in ancient India.

Following The Bronze Age was The Iron Age. Extracting iron from its ore into a metal was more difficult than copper and tin. Iron was not better suited for tools than bronze until steel was actually discovered. Iron ore was considered to be much more abundant and common than either copper or tin. Iron appeared to have been invented by The Hittites around 1200 BC which began The Iron Age. Extracting and working iron was a key factor for the success of the Philistines. We can think of The Iron Age as the advent of iron working or ferrous metallurgy. There have been historical developments in ferrous metallurgy and found in a wide variety of past cultures and civilizations. They include ancient and medieval kingdoms and empires of The Middle

East and Near East, ancient Iran, ancient Egypt, ancient Nubia, Anatolia (Turkey), Ancient Non, Carthage, the Greeks, and Romans of ancient Europe, medieval Europe, ancient and medieval China, ancient and medieval India, ancient and medieval Japan, and many others.

As years progressed during these antiquity ages, philosophers contributed there expertise to the fields of science. There were philosophical attempts to rationalize why there were different substances that have different properties such as color, density, and smell that exist in different states such as gaseous, liquid, and solid, and that these substances react in different manners when exposed to environments like water, fire, or temperature changes that led ancient philosophers to postulate the first theories on both nature and chemistry. These philosophical theories that relate to chemistry have been traced back to every single ancient civilization.

The common aspect of these theories was the attempt to identify a small number of classical elements that make up various substances in nature. Substances such as air, water, and soil/earth, and energy forms such as fire and light and abstract concepts such as thoughts ether and heaven were common in ancient civilizations. Ancient Greek, Indian, Mayan, and Chinese philosophies considered air, water, earth, and fire to be the primary elements.

During the ancient world around 420 BC, Empedocles stated that matter was made up of four elemental substances which were earth, fire, air, and water. Early theories of atomism can be traced back to ancient Greece and ancient India. Greek atomism was popular by the Greek philosopher Democritus who declared mater to be composed of indivisible and indestructible particles he called "atomos". This took place around 380 BC. Earlier, Leucippus stated that atoms were indivisible parts of matter. This coincided by the Indian philosopher Kanada in his Vaisheshika sutras around the same time period. Aristotle was different because he opposed the existence of atoms in 330 BC. Others had their own theories such as Poly's the physician around 380 BC. His Greek text argued that the human body was made up of four humors instead. Epicurus around 300 BC postulated a universe made up of indestructible atoms in which we are responsible for achieving a balanced life.

The Roman poet and philosopher Lucretius had a goal of explaining Epicurean philosophy to the Roman audience. He wrote De rerun nature

"The Nature of Things" in 50 BCE. In this work, Lucretius presents the principles of atomism; the nature of the mind and soul; explanations of sensation and thought; the development of the world; and an explanation of celestial and terrestrial matters.

It was also the earliest alchemists in the Western tradition that have came from GrecoRoman Egypt in the first centuries AD. They invented technical work and invented chemical apparatuses. In fact, the bain-marie or water bath is named for Mary the Jewess. Her work gave the first descriptions of the tribikos and kerotakis. Cleopatra the Alchemist described furnaces and was credited with the invention of the alembic. Later on, Zosimos of Panopolis decided to write books on alchemy which he called cheirokmeta which was the Greek word for "things made by hand". These works included references to recipes and procedures and descriptions of instruments. Early development of purification methods were described by Pliny the Elder in his Naturalist History. He did his best to explain these methods and make acute observations of the state of several minerals.

As more years progressed, medieval alchemy became popular. An elemental system used in medieval alchemy was developed by the Persian or Arab alchemist and pioneer of organic chemistry Jabir bin Hayyan, and it was based on the classical elements of Greek tradition. His system consisted of four Aristotelian elements of air, earth, fire, and water with the addition of two philosophical elements. These elements were sulphur which characterizes the principle of combustibility "the stone which burns" and mercury which characterizes the principle of metallic properties. They were seen by early alchemists as expressions of irreducible components of the universe and were a larger consideration within philosophical alchemy.

The three metallic principles such as sulphur to flammability or combustion, mercury to volatility and stability, and salt to solidity became the tria prima of the Swiss alchemist Paracelsus. Paracelsus reasoned that Aristotle's four-element theory appeared in bodies as these three principles. He saw the principles as fundamental and justified them to the description of how wood burns in fire. Mercury included the cohesive principle so when it left the wood in smoke the wood fell apart. Smoke described the volatility which was known as the mercurial principle, the heat-giving flames described flammability was known as sulphur, and the remnant ash described solidity known as salt.

Overtime, alchemy was defined by the Hermetic quest for the philosopher's stone which was the study of symbolic mysticism and that differed greatly from modern science. Alchemists toiled to make transformations of esoteric or spiritual and/or exoteric or practical level. It was the protoscientific and esoteric aspects of alchemy that contributed to the evolution of chemistry in Greco-Roman Egypt, in the Islamic Golden Age, and then in Europe. Both alchemy and chemistry share an interest in the composition and properties of matter. Even until the 18th century they were not separate disciplines. It was the term chemistry used to describe the blend of alchemy and chemistry that existed before that time period. The Renaissance Period was different and during this time period esoteric alchemy was

popular in the form of Paracelsian iatrochemistry, while spiritual alchemy flourished and realigned its Platonic, Hermetic, and Gnostic roots. The symbolic quest for the philosopher's stone was not superseded by scientific advances. Many respected scientists and doctors still held onto these ideas until the early 18th century. Jan Baptist van Helmont, Robert Boyle, and Isaac Newton were east modern alchemists and were renowned for their scientific contributions.

We have also talked about alchemy in the Islamic world. The Muslims took the time to translate the works of ancient Greek and Hellenistic philosophers into Arabic and then were experimenting with scientific ideas. The 8th-century alchemist Jabir bin Hayyan introduced a systematic classification of chemical substances and also provided instructions for deriving an inorganic compound such as sal ammoniac or ammonium chloride from organic substances such as plants, blood, and hair by chemical methods. There were some Arabic Jabirian words such as the "Book of Mercy" and the "Book of Seventy" which were translated into Latin under the latinized name "Geber". During the 13th century in Europe, there was an anonymous writer referred to as pseudoGeber who produced alchemical and metallurgical writings under his name. There were still other influential Muslim philosophers such as Abu al-Rayhan al-Biruni and Avicenna who disputed the theories of alchemy which was the theory of the transmutation of metals. But as years progressed, the fields of alchemy and intra-chemistry changed. The birth of modern chemistry probably took place with the scientist Robert Boyle. He recognized its aim was neither the transmutation of metals nor the preparation of medicines, but instead was all about observation and generalization of certain classes of phenomena. He did not agree the validity of the views of the alchemists based on matter. He came up with the idea and definition of an element which has reigned supremacy in chemistry. He also enriched the sciences with observations of importance. Robert Boyle had ideas that were advanced of his time intervening between the iatrochemical period and the Age of Modern Chemistry which came a period of time called The Phlogistic Theory. This theory had certain connections with the ideas of the alchemists. The phlogiston theory was due to Georg Ernst Stahl during the late 1600s and early 1700s.

Becher had attempted to revive and modified the sulphur-mercury-salt theory of alchemists. He came up with the assumptions that all substances consist of three earths such as the combustible, mercurial, and vitreous. This was found to be the germ of Stahl's phlogistic theory. Based on Stahl, he felt that all combustible bodies including metals that change on heating have phlogiston. It was the principle of combustion that escaped in the form a flame when substances were burned. According to the theory, metals were compounds since they have metallic calx which we now call the oxide of the metal combined with phlogiston. In order to obtain the metal from the calx, it was necessary to act upon it the same substance that was rich in phlogiston.

Coal and charcoal were easily combustible and leaves very little residue. According to this theory, they consist largely of phlogiston and metals that can be obtained by heating their calces with either of these substances. Many other facts in nature were similar to the phlogiston theory. Chemists during this time period did not pay much attention to the balance. It was observed that metals increased in weight on calcination, but this explained on the assumption that phlogiston had negative weight. It was not until the scientist Antoine Lavoisier during the middle of the 1700s who utilized Priestley's discovery of oxygen which he called to be dephlogisticated air and also studied the weight relations which accompany combustion who demonstrated how nonvalid the phlogistic theory was. He proved that combustion of a substance burned with a certain constituent of air he called oxygen. It was not until this time that alchemy was defunct. It was Boerhave who was the last eminent chemist to support any doctrines, and the new chemistry of Lavoisier came about. This was the time period we enter "The Age of Modern Chemistry".

Remember it was Robert Boyle who had defined an element as a substance that could not be decomposed but entered into combination with other elements to give compounds capable of decomposition to the original elements. The metals were classed as the elements since they defied attempts to decompose them. The definition had a negative character. It was better to term elements as all substances that has defied decomposition. It was a matter of impossibility to decide what substances were true elements and that gold and other metals were of a compound nature as well as the possibility of preparing gold from base metals or other substances. The uncertainty regarding the elements have been recognized by these chemists.

The study of relative weights for which substances combine were discovered. The stoichiometric laws were then proposed. The Law of Constant Proportion states the same chemical compound always has the same elements, and there is a constant ratio between the weights of elements present. The Law of Multiple Proportions states that if two substances combine chemically in more than one proportion, then the weights of one which combine with the weights of the other are in simple ratios. The Law of Combining Weights state that substances combine either in the ratio of combining numbers or in simple rational multiples or submultiples of the numbers. We think of it as the weights of different substances that combine with a given weight of some particular substance to be taken as a unit. They are called combining numbers of substances with reference to the unit.

The usual unit during this time period was eight grams of oxygen.

Based on these laws, we can state the following facts. Pure water was found to always have hydrogen and oxygen combined in the ratio of 1.008 parts by weight to 8 parts by weight. As another example, pure sulphurdioxide was always found to consist of sulphur and oxygen combined in the ratio 8.02 parts by weight of sulphur to 8 parts by weight of oxygen. This was what we refer to as The Law of Constant Proportions. Yet we find another compound that consists of only oxygen and hydrogen which differs from the properties of water. It was also found that hydrogen and oxygen combined in the ratio of 1.008 parts by weight to 16 parts by weight.. The weight of hydrogen was combined with the amount of oxygen to be twice combined with the same weight of hydrogen in water. There has not been one definite compound that has been discovered with an intermediate between the two. There were other compounds that consist of sulphur and oxygen. As examples, sulphur-trioxide or sulphuric anhydride was found to consist of sulphur and oxygen combined in the ratio of 5.35 parts by weight of sulphur to 8 parts by weight of oxygen. The weights of sulphur combined with a definite weight of oxygen in the two compounds were called sulphurdioxide and sulphur-trioxide being in the proportion of 8.02 to 5.35 and there were similar simple ratios being obtained for other compounds. This is what we call The Law of Multiple Proportions.

Based on the data above, we can fix the number of hydrogen as 1.008 to that of sulphur to be 8.02. Here we have compounds that contain both sulphur and hydrogen. The weight of sulphur combined with 1.008 grams of hydrogen was always found to be 8.02 grams or some multiple or submultiple of the quantity. In a simple compound containing hydrogen and sulphur, there was 1.008 grams of hydrogen to be found always combined with 16.04 grams of sulphur which was exactly twice the quantity above. This was referred to as The Law of Combining Weights. Berthollet did not agree about the truth of The Law of Constant Proportion. There was a controversy between Berthollet and Proust. Joseph Proust carried out his research and felt his results were in agreement with The Law of Constant Proportions.

Towards the beginning of the nineteenth century, John Dalton came up with his idea of The Atomic Theory. His theory assumes that all matter is made up of small indivisible and indestructible particles which he referred to as atoms, that all atoms are not the same meaning there are many sorts of atoms as there are elements, that atoms constituting one element is exactly alike and of definite weight, and that compounds can be produced by combining different atoms. If matter is so constituted, then the stoichiometric laws must follow. The smallest particle of a definite compound which is now called a molecule consist of a definite assemblage of different atoms and the atoms have definite weight which still agrees with The Law of Constant Proportion. What we know about The Law of Multiple Proportions is when one atom of one substance combines with 1,2,3.....atoms of other substances, it cannot combine with fractional parts of the atom since the atoms are indivisible. Since the atoms have definite weights, then The Law of Combining Weights follow. Dalton's Atomic Theory gave a simple explanation of the facts regarding the weights of substances entering into chemical combination and gained universal acceptance. But there was still a spirit of revolt against it as an absolute constitution of matter. The tendency of scientific philosophy has been towards monism as opposed to dualism.

Dalton's theory did not agree with the unity of the Cosmos. It lacked the unifying principle of alchemists. It was only in recent times that it has been recognized that a scientific hypothesis might be useful without being altogether true. Yet Dalton's theory can be accomplished which no other hypothesis could have been with during this time period. It rendered the concepts of what we call chemical elements, a chemical compound, and a chemical reaction being definite. It has also led to the majority of discoveries in chemistry. But still his theory fails to be 100% satisfactory. The philosophers of materialistic school of thought who insisted the absolute truth of The Atomic Theory. Kekule developed Franklin's theory of atomicity or valency made more definite the atomic view of matter. Kekule had doubts to the truths of Dalton's theory, but he regarded as true. But he still regarded it as chemically true that there were such things as chemical atoms and chemical elements that were not capable of being decomposed by purely chemical means and they were not absolute atoms or absolute elements.

Eventually Dalton's Atomic Theory was accepted, and it was necessary to determine atomic weights of various elements. It was the relative weights of various atoms with reference to one of them as a unit. The correct atomic weights came about due to the acceptance of Avogadro's Hypothesis. His hypothesis stated that equal volumes of different gases measured at the same temperature and pressure contain an equal number of gaseous molecules was put together due to the physical behavior of gases. Its importance was not recognized due to some time. The distinction between atoms and molecules were not drawn.

It was not until the observation of a chemist by the name of Prout that the atomic weight of hydrogen was taken as the unit and the atomic weights of nearly all elements appear to be whole numbers. In the year 1815, he suggested that all elements consist solely of hydrogen. Prout's Hypothesis received favorable reception. It harmonized Dalton's Theory with the concept of unity of matter and that matter was hydrogen in essence.

Thomas Thomson did his research to

demonstrate the truth. The eminent Swedish chemist Berzelius who carried out atomic weight determinations criticized Prout's Hypothesis and Thomson's research. He felt that the decimals in his atomic weights obtained experimentally by Berzelius had so many errors. In the year 1844, Marignac suggested half the hydrogen atom as the unit since the element chlorine with an atomic weight of 35.5 would not fit with Prout's Hypothesis. Even later on Dumas suggested one-quarter. With the theoretical-division of the hydrogen atom, the hypothesis was found to be more complicated. We know that recent and most accurate atomic weight determinations show that the atomic weights were not whole numbers but the majority of them approximate to such. The Hon. R.J. Strutt has calculated the probability of this taking place with common elements by mere chance to be exceedingly small. There were modern scientific speculations that concern the constitution of atoms towards a modified form of Prout's Hypothesis or to view the atoms of other elements that were polmerides of hydrogen and helium atoms. According to modern views, it was possible for elements of different atomic weights to have identical properties since they depend on the number of free electrons in the atom and not on the massive central nucleus. Sir Joseph Thomson and Dr. F.W. Aston discovered the element neon was a mixture of two isotopic elements having unequal proportions where one having an atomic mass of 20 and another atomic mass of 22. Dr. Aston has perfected the method of analyzing mixtures of isotopes and also determined their atomic masses. The atomic weight of hydrogen to be 1.008 was confirmed. The elements helium, carbon, nitrogen, oxygen, fluorine, phosphorus, sulphur, arsenic, iodine, and sodium were found to be simple bodies with whole-number atomic weights. However, boron, neon, silicon, chlorine, bromine, krypton, xenon, mercury, lithium, potassium, and rubidium were found to be mixtures. The indicated atomic mass of each of the constituents were whole numbers. As an example, chlorine, who has an atomic weight of 35.46, was found to be a mixture of two chemically-identical elements which were 35 and 37. Other elements such as xenon are mixtures of more than two isotopes. It was highly probable that what was true of the elements was probably true for the remainder. It appeared the irregularity presented by the atomic weights of ordinary elements have confused scientists in the past due to the fact that these elements were mixtures. As for hydrogen, it was only reasonable to suppose the close packing of electrically charged particles should give rise to a slight decrease in the total mass so then the atomic weights of other elements as for hydrogen being equal to one should be slightly less than the whole numbers or the atomic weight of oxygen referred to as sixteen should be slightly more than unity.

As years progressed, properties of atomic weights were discovered independently by Lothar Meyer and Mendeleeff. They found the elements could be arranged in rows of their atomic weights so similar elements would be found in the same columns. The alkali metals such as lithium, sodium, rubidium, and caesium which resemble each other are found in the first column. The alkaline earth metal s are found in the second column even though some of the elements had different properties. They find on the table remarkable regularity with some divergences. This shows the properties of the elements are periodic functions of their atomic weights. Properties of the elements distinguishes them from compounds. It was probably concluded with tolerable certainty that if the elements were a compound of nature, then they were a compound of like nature distinct from other compounds.

It was not until the late Sir William Crookes attempted to explain the periodicity of the properties of elements on the theory that they have been evolving by a conglomerating process from primal stuff referred to as prostyle which consisted of small particles. He represented the action of this cause by means of a “figure of eight” spiral along which the elements were placed at regular intervals so similar elements come underneath one another based on Medeleeff’s table. It was found the slope of the curve was supposed to represent the decline of some factor such as temperature conditioning the process which recurs in nature like the swing of a pendulum. After the completion of a swing where one series of elements was produced, then the same series of elements was not again the result as would be the case, but a different series was produced where each member resembled the corresponding member of the former series. As an example, if the first series contains helium, lithium, carbon, etc.., then the second series contain argon, potassium, titanium etc.. The whole theory is free from defects.

We are going to turn our attention to recent views of matter which originated to a great extent in the investigations of the passage of electricity through gases at very low pressures. This is what we refer to as "The Corpuscular Theory of Matter". When an electrical discharge has passed through a high-vacuum tube, there are invisible rays being emitted from the cathode with the production of greenish-yellow fluorescence where they strike the glass walls of a tube. They were called "kathode rays". During this time period, they were regarded as waves in the ether, but it was shown that Sir Williams Crookes that they consist of small electrically charged particles that move with high velocity. Sir J.J. Thomson was able to determine the ratio of the charge carried out by the particles to the mass or inertia. He found the ratio to be constant regardless of the gas contained in the vacuum tube and there was a much greater than the corresponding ratio for the hydrogen ion in electrolysis. Mr. C.T.R. Wilson discovered that charged particles might serve as nuclei for the condensation of water-vapor. He was able to figure out the value of the electrical charge carried out by these particles. It was found to be constant and equal to the charge carried by univalent ions such as hydrogen in electrolysis. The mass of the cathode particles should be much smaller than the hydrogen ion. The actual ratio was found to be 1:1700. The first theory that was put forward by Sir J.J. Thomson was that cathode particles, which he referred to as corpuscles, were electrically charged portions of matter. They were much smaller than the smallest atom. The same sort of corpuscle was obtained regardless of the gas contained in the vacuum tube. They assumed and concluded the corpuscle was the common unit of all matter.

This eminent physicist showed mathematically that a charged particle which moves with high velocity approaching that of light would exhibit an increase in mass or inertia due to charge which is the magnitude of such inertia that depends on the velocity of the particle. This was verified by Kaufmann who determined that velocities and ratios between electrical charge and inertia of various cathode particles and similar particles emitted by compounds of radium. Sir J.J. Thomson calculated the values assuming the inertia of such particles was of electrical origin. He obtained values in agreement with experimental values. There was no reason for supposing the corpuscle to be matter after all. If this was the case, then the above agreement should not have been obtained. It was thought that the corpuscle was nothing but a disembodied electrical charge which contained nothing material. It was electricity and nothing but electricity. A new term was then coined and it was the term electron. We have what we call "The Electronic Theory" which was the material atoms consisting of electrons or units of electricity in rapid motion. Matter was simply an electrical phenomenon.

Sir J.J. Thomson elaborated his theory of nature and constitution of matter. He showed which electrons would be stable and attempted to find the significance of Mendeleeff's generalizations and explanations of valency. This was considerable truth to the electronic theory of matter. In other words, inertia can be accounted for electrically. There was still that difficulty that electrons were units of negative electricity where matter was electrically neutral. There have still been theories that has surmounted this difficulty. The electron is a constituent of matter. Based on recent research, all atoms consist of two distinct portions which is a massive central nucleus

whose net charge is positive and being surrounded by a number of electrons to neutralize the charge. The indicated number of free electrons is exactly the number that expresses the position of the elements on the periodic table which made sense since helium would be two and lithium would be three. It seems that the chemical properties of elements have been determined entirely by electrons and are periodic functions of their atomic weights as was thought but instead were there atomic numbers. The exact nature of the nuclei of various atoms had to be determined. It was thought the case of the atoms being heavier than helium would appear to be made up of nuclei of hydrogen or helium atoms being together with electrons insufficient in number to neutralize positive charges.

The analysis of matter was then carried out a step further. There is still that philosophical view of the cosmos that involves the assumption of absolutely continuous and homogeneous medium which fills all space, an absolute vacuum is not thinkable, and it was supposed that the stuff filling the space is of atomic nature. They were curious as what occupied the interstices between atoms. The ubiquitous medium was termed "the Ether of Space". The ether is demanded by light. It appeared the ether of space had another more important function than the transmission of light. Matter had its explanations developed by Sir Oliver Lodge. The evidence points to the conclusion that matter was a sort of singularity in the ether which was probably a stress center. We have become accustomed to think the ether as excessive light and the reverse of massive or dense in which we were wrong. Sir Oliver Lodge calculated the density of ether being far greater than most dense forms of matter. It is not that matter is to be thought of being a rarefaction of the ether, for the ether in matter is as dense even without it. What was called matter was not a continuous substance. It consisted of widely separated particles, and its small density compared to perfectly continuous ether. If there was a difficulty to think how a perfect liquid like ether gave rise to a solid body possessed by properties as rigidity, impenetrability, and elasticity, then we must remember that the properties can be produced by motion. It appeared that the ancient doctrine of alchemistic essence was fundamentally true after all. There was out of the "One Thing" all materials have been produced by both adaptation and modification. There seemed to be some resemblance between the concept of the electron and the seed of gold where the seed was regarded by alchemists as a common seed of all metals.

There still needs to be some modification of Dalton's Atomic Theory as found in The Electronic Theory. Each chemical element gives a spectrum due to its atoms at a sufficiently high temperature to act upon the ether to produce vibrations of definite wavelengths. Each spectrum would be considered to be different for each element. It is incredible that an atom would give rise to a number of different vibrations and the only conclusion that could be found was the complexity of the structure. Spectroscopic examination of heavenly bodies leads to the conclusion that there is a process of evolution at work that builds up complex elements from simpler ones. This makes sense since the hottest nebulae appeared to consist of a few simple elements while cooler bodies exhibit greater complexity.

Atoms can no longer be viewed as eternal and indissoluble gods of nature that they were once supposed to be. Materialism was deprived of what was thought to be a scientific basis. The science of chemistry was not affected and the atoms were not the ultimate units for which material things were built. Atoms cannot be decomposed by purely chemical means. The elements really were not truly elemental, but were considered to be chemical elements. The atomic theory had been a subject of criticism. Wald argues that substances should obey The Law of Definite Proportions. This was the way they should be prepared. Chemists refused to admit any substance as a definite chemical compound unless it obeys The Law of Definite Proportions. It was Wald's conjectures that was supported by Professor Ostwald who attempted to deduce other stoichiometric laws without assuming any atomic hypothesis. But the ideas do not appear to have gained the approval of chemists. Chemists still accept Dalton's Atomic Theory as a starting point.

There still seems to be logic for the arguments of Wald and Ostwald, but still the direction of scientific theory and research did not appear in Wald's views. The atomic theory was not necessitated by stoichiometric laws. There seemed to be a molecular constitution of matter to be known as "Brownian movement" which was the spontaneous irregular and perpetual movement of microscopic portions of solid matter when it was immersed in a liquid medium. This movement appeared only to be the result of motion of the molecules for which the liquid has built up.

We now come to the final stages where are going to turn our attention to modern alchemy, but technically there is no such thing as modern alchemy. I'm not saying that mysticism is dead or that men don't seek to apply the principles of mysticism to phenomena on the physical plane, but they still do so after another manner from alchemists. We can consider this be a new science related to both chemistry and physics but more profound and more detailed. It is this science which we should use the expression "modern alchemy". We've learned in physics that X-rays are produced when there is an electrical discharge being passed through a high-vacuum tube. The rays are a series of irregular pulses in the ether when the cathode particles strike the walls of the glass vacuum tube. More powerful effects are produced by inserting a disc of platinum in the path of kathode particles. It was the scientist by the name of M. Becquerel who discovered that these substances emit radiations that are similar to X rays. He found uranium compounds that were affected by a photographic plate and from which they were screened showed uranium radiations or "Becquerel rays" resembling X-rays. It was known that certain substances fluoresce or emit light in the dark after being exposed to sunlight. It was once thought uranium salts would exhibit this phenomenon since certain uranium salts fluorescent. M. Becquerel found uranium salts which had not been exposed to sunlight affected a photographic plate regardless if it fluorescent or not. The phenomenon is called radioactivity. Bodies that exhibit it are known to be radioactive. Another scientist by the name of Schmidt found thorium compounds to possess a similar property, and Professor Rutherford showed thorium compounds to evolve resembling a gas and called this “emanation”.

Madame Curie determined the radioactivity of many of the uranium and thorium compounds. She found there to be a proportion between radioactivity of compounds and quantity of uranium or thorium in them with the exception of certain natural ores. Some certain natural ores had a radioactivity greater than pure uranium. Madame Curie prepared one of the ores by a chemical process and found it possessed normal radioactivity. It was only logical that the ores contain some unknown highly radioactive substance. The Curies were able to extract pitchblende which had the ore of great radioactivity. But the extraction of pitchblende was only minute quantities of the salts of two new elements, and they were named "polonium" and "radium". Both of these were extremely radioactive. The scientist M. Debierne obtained a third radioactive substance from pitchblende and called it "actinium". Radium resembles calcium, strontium, and barium in chemical properties. The atomic weight was determined by Madame Curie and was found to be about 225 based on her first experiments. The scientist Sir T.E. Thorpe determined the value to be a little bit higher than 225. Radium gave a characteristic spectrum and was really radioactive. Towards the middle of the year 1910, the element radium had not been prepared. Compounds such as radium chloride and radium bromide have been employed. Madame Curie and M. Debierne obtained the free metal. It was described as a white shining metal resembling the other alkaline earth metals. It reacts with water but in a violent manner, chars paper, and blackens the air which owes to the formation of nitride. It has the ability to fuse at 700°C, and it was more volatile than the element barium.

Radium salts have the ability to give three rays which we can denote as alpha, beta, and gamma. Alpha rays consist of electrically charged positive particles with a mass equal to that of four hydrogen atoms. They can become deviated by the magnetic field and don't possess great penetrative power. Beta rays are similar to kathode rays and consist of negative electrons. They are also deviated from a magnetic field but opposite in direction and possess medium penetrative power that passes through a thin sheet of metal. Gamma rays resemble X-rays and possess great penetrating power and are not deviated from a magnetic field. Differences in the effect of the magnetic field and differences in their penetrative power lead to their detection and allowed separate examinations.

Radium salts emit an emanation and tended to be occluded in the solid salt, but it can be liberated by dissolving the salt in water or by heating it. The emanation exhibits characteristics properties of a gas. In other words, it obey's Boyle's Law where volume varies inversely with pressure. It can be condensed to a liquid at low temperatures. Its density by the diffusion method is about 100. Attempts to prepare chemical compounds of emanation have failed. This resembles rare gases in the atmosphere such as helium, neon, argon, krypton, and xenon where it is likely these molecules are monatomic so the density of 100 would give the atomic weight of 200. Atomic weights of about 220 corresponds to the position of the column that contains rare gases. The emanation that has an atomic weight of these dimensions have been confirmed by further experiments carried out by late Sir William Ramsay and Dr. R. W. Gray. It was these two chemists that determined the density of the emanation by weighing minute quantities of known volume of the substance sealed in a small capillary tube with a balanced being employed. They found the values for the density to range from 108 to 113.5 which corresponded to values for the atomic weight varying from 216 to 227. Sir William Ramsay considered that the emanation had to be one of the elements of the group of chemically inert gases. He proposed the name "Niton" and figured it had a probability of an atomic weight of about 222.5.

Radium salts also have the property by emitting light and heat. It seems contradictory to The Law of Conservation of Energy. But it does become clear in terms of the corpuscular or the electronic theory of matter. It was thought that the radium-atom to be a system of a large number of corpuscles or electrons and contains an enormous amount of energy. It contains atomic systems such as molecules that contain much energy to be unstable and able to explode. The radium atom does have the ability to explode, and energy is set free and manifests itself as heat and light. Some of the free electrons get shot off which are the beta-rays and strike the undecomposed particles of salt and give rise to pulses in the ether which were the gamma rays. It was similar to kathode particles that give rise to Xrays when they strike walls of a vacuum tube or a platinum disc that was placed in their path. The beta and gamma rays do not result immediately from the radium atoms exploding. The initial products were from the emanation and one alpha particle from each radium atom destroyed.

Radium salts cause surrounding objects to become radioactive for a short amount of time. The "induced radioactivity" was due to the emanation which was radioactive and only emitted alpha rays. It was decomposed into minute traces of solid radioactive deposits. The rate of decay of the activity of the deposit undergoes a series of subatomic changes where the products were being termed Radium A, B, and C. Beta and gamma rays emitted by radium salts were due to secondary products. Radium F was thought to be identical to Polonium and there were still other products formed. The first product of uranium and thorium decompose to the first product as a solid.

Sir William Crookes was able to separate from uranium salts a small quantity of a radioactive substance he called Uranium X. The residual uranium lost most of the activity. M. Becquerel repeated the experiments and found the activity of the residual uranium was slowly regaining while Uranium X decayed. He came up with the theory that uranium first changes to Uranium X. He suggested that radium might be the final product breaking up the uranium atom. It was certain uranium that would be evolved and it would all decompose. This has been experimentally confirmed due to the growth of radium being in large quantities of a solution of purified uranyl nitrate. Uranium does not give emanation. The element thorium gave at least three products such as meso-thorium, radio-thorium, and Thorium X. Thorium X yields an emanation that resembles radium but it was not identical to it.

The Radium Emanation method was unique. If we distill off the emanation of some radium bromide and measure the quantities of heat given off, then both Rutherford and Barnes proved that three-fourths of the total amount of heat given by a radium salt comes from minute quantity of emanation. The amount of energy liberated as heat by the series of decay of this emanation was huge. It only takes one cubic centimeter to liberate about four million times as much heat as obtained by combusting an equal volume of hydrogen. It was observed that radioactive minerals on heating release helium which was a gaseous element. It was very likely that helium was an ultimate product during decomposition. Sir William Ramsay and Mr. Soddy carried out careful experiments by using a U-tube cooled in liquid air and then extracted by pumps. They found a helium spectrum to be produced. Sir William Ramsay carried out a further experiment where the radium salt had been heated in a vacuum for some time. It was proven the helium could not be occluded.

Sir James Dewar confirmed Sir William Ramsay's results. He suggested the alpha particle consist of a electrically charged helium-atom. The view is in agreement with the mass of the particle determined experimentally, but it was demonstrated by Rutherford and Royds. They carried out an experiment in which the emanation of about one-seventh of a gram of radium was enclosed in a thin-walled tube through the walls where alpha particles could pass but at the same time impervious to gases. The tube was surrounded by an outer jacket that was evacuated. The presence of helium in the space between the inner tube and outer jacket has been observed by spectroscopic means. The emanation atom results from the radiumatom by getting rid of one alpha particle and the latter consists of an electrically charged helium atom. It followed that the emanation have an atomic weight of 226-4. The value was in agreement with Sir William Ramsay's determination of the density of emanation. The formation of one element into another took place, and it was proven that radium was a chemical element. It was suggested that radium might be composed of helium with some unknown element or a compound of helium with lead. It has been shown that lead was probably one of the end products for the decomposition of radium. The following suggestions came about:

i)Attempts to prepare compounds of helium with other elements have not worked out ii)Radium has properties of an element, has a spectrum, and is included with similar properties with other elements on the periodic table.

iii)Quantity of heat liberated on the decomposition of emanation is observed

iv) The rate of decay of the emanation is not affected by changes of temperature where chemical actions are affected in rate by changes due to temperature.

There were also differences between the helium and emanation. The latter was considered to be a heavy gas that was condensable to a liquid by liquid air, whereas helium was the lightest of all known gases except hydrogen. Helium has also been liquified by persistent effort. The emanation was radioactive which gives off alpha particles. Helium does not have this property. It was pointed out that the change was not meant by the term "transmutation". We consider it to be a spontaneous change. It still does not prove that the chemical elements were not in discrete units of matter. All matter was radioactive to a slight degree. They felt the chemical elements were not permanent, and they undergo change. Common elements mark the points where the rate of evolutionary process is slowest. We can now disprove the old alchemistic doctrine that metals grow in the womb of nature.

We have mentioned earlier that radium emanation contained a store of potential energy. The idea of utilizing this energy brought about chemical changes according to Sir William Ramsay. He did his research on the chemical action of the substance with surprising and interesting results. The energy contained within the radium emanation acted like The Philosopher's Stone. First experiments were carried out on distilled water. The emanation decomposed water into gaseous elements oxygen and hydrogen where hydrogen was produced in excess. The results were confirmed and hydrogen peroxide was detected which explained the formation of excess hydrogen. They also showed the emanation brought about a reverse change causing oxygen and hydrogen to combine to form water until a position of equilibrium was observed. The gas obtained by the action of the emanation on water after removing ordinary gases gave a surprising result. The gas showed a spectrum of neon with some faint helium lines. Sir William Ramsay and Mr. Cameron carried out experiments with a silica bulb instead of glass. A spectrum of the residual gas after removing ordinary gases was photographed. There was a large number of neon lines that were identified. Helium was also present. Ramsay felt the presence of neon was not obtained by the leakage of air into the apparatus. The percentage of neon in the air was not sufficiently high. The suggestion may have put forward the element argon. They also felt the neon could not have come from the aluminum of the electrodes as a sparking tube has been used and tested before an experiment was carried out. Rutherford and Royds were not able to confirm the results. They described attempts to get neon by emanation.

Out of five experiments, none of the neon was obtained. There was a small air leak that was discovered. Minute quantities of the gas were sufficient to give a visible spectrum. Ramsay's results were due to leakage of air into the apparatus. But if this was the true explanation of Ramsay's results, it was difficult to know why an experiment with a solution of a copper salt had not been detected if due to leakage the proportions of rare gases should have been the same in all experiments. An excess hydrogen was produced when water decomposed by the emanation which was suggested to Sir William Ramsay and also Mr. Cameron. They felt if a solution of a metallic salt had been employed in place of water, then they felt the free metal would be obtained. There were modern alchemists that investigated the action of radium emanation of copper solutions and lead salts and again transmutations taking place. They found by removing copper from a solution of a copper-salt, there was a considerable quantity of sodium present together with traces of lithium. The gas evolved and contained along with nitric oxide and a little nitrogen and argon but there was no helium. It seemed like a dual transmutation copper into lithium and sodium and then an emanation to argon. They also found out that carbon dioxide evolved from an acid solution of thorium nitrate. In fact, helium, neon, and argon occur in the same column on the periodic table of elements with emanation whereas lithium and sodium went with copper and carbon went with thorium. The elements being produced lighter atomic weights than those decomposed.

Later on, Madame Curie and Mademoiselle Gleditsch decided to repeat Cameron and Ramsay's experiments on copper salts using a platinum apparatus. They did not detect lithium after the action of emanation. Cameron and Ramsay's results might be due to the glass vessels employed. Dr. Perman investigated the action of emanation on copper and gold and even failed to detect any traces of lithium. The transmutation of copper to lithium was still not proved, and there needed to be further research. Sir William Ramsay gave a presidential address to the Chemical Society on March 25, 1909 after having brought forth arguments for the possibility of transmutation. He described experiments he carried out on thorium and similar elements. It was found that carbon dioxide was evolved from an acid solution of thorium nitrate. It also appeared carbon dioxide was produced by radium emanation on thorium nitrate. The action of radium emanation of other members such as silicon, zirconium, and lead was carried out. When zirconium nitrate and hydrofluosilicic acid was used, it was found that carbon dioxide was obtained, but using lead chlorate the amount of carbon dioxide was not significant. The perchlorate of bismuth

also gave carbon dioxide by emanation. Sir William Ramsay and Mr. Francis L. Usher carried out experiments with a compound of titanium. The results confirmed their former experiments. When lead was used, the amount of carbon dioxide was not appreciable.

A suggestive argument for the transmutation of metals was put forth by Professor Henry M. How einhis paper "All otropyor Transmutation". Certain substances were known which differed in their physical properties behave chemically as if they were one and the same element which gave rise to the same series of compounds. Typical substances include diamond, graphite, and charcoal all to contain carbon just like yellow phosphorus and red phosphorus being convertible to one another. The substances were were different forms or allotropic modifications of the same element and not to regard them as being different elements. Based on atomic theory, the differences between "allotropic modifications" were ascribed to differences in the number and arrangement of atoms constituting molecules of "modifications". Yet we cannot argue that these modifications which are transmutable into one another are the same element just because they have the same atomic weight. Different elements were distinguished by different atomic weights. In the determination of atomic weights, derivatives of bodies are employed. The value obtained was the atomic weight of the element which formed derivatives. The modifications were different since equal weights contain different quantities of energy. The change of one form to another form takes place with the evolution of heat. According to modern views on the nature of matter, this was the sole fundamental difference between two different elements since equal weights contain equivalent to different quantities of energy. These so-called "allotropic modifications" of elements were just as much different elements as any other different elements and the change from one modification to another was a true transmutation of the elements. There was only one distinction and that they differ in respect of the energy contained. They were easily able to convert one into the other, whereas different elements differed greatly from one another. The transmutation of one such element into another element was obtained by the utilization of energy in a highly concentrated form such as those that were evolved simultaneously with the spontaneous decomposition of radium emanation. What we have shown up to this point was modern science indicating the essential truth of alchemistic doctrine.

Whatever may be the final analysis of Sir William Ramsey, those of Sir Ernest Rutherford demonstrated the facts of transmutation. It is worth noting how many of the alchemists obscure descriptions of their magic and sorcery apply to what we refer to as energy which is the true "First Matter" of the universe. The beginnings of realistic modern chemistry with the discovery of more detailed structures of the atom was soon on its way during the early 1900s.

ABOUT THE AUTHOR

Harminder Gill did his undergraduate studies at the University of California, Riverside where he finished his Bachelor of Science Degree in Biological Chemistry with an Emphasis in Chemistry. After completing his undergraduate studies, he attended graduate school at the University of California, Davis, where he finished his Master's Degree in Organic Chemistry. After doing several teaching assignments, he went into the teaching profession at community colleges in Southern California. He became a life member for both the alumni associations at UCR and UCD. He has done homeschooling throughout communities, where he offers more than two hundred sections of academic subjects from humanities, social sciences, sciences, and test preparation. Plenty of his past students have done exceptionally well, including those who have achieved high scores on standardized tests, getting accepted into training academies, universities, and programs of their choice. He has served on Board of Directors for an art association in Corona, and he was served on the Board of Library Trustees Ward 4 in the city of Riverside. He has also completed the Distinguished Toastmaster (DTM) in the Toastmasters Program for Public Speaking and Leadership, and he does his best to keep current in the math, sciences, and technology.

REFERENCES

Asimov, Isaac. A Short History of Chemistry An Introduction to the Ideas and Concepts of Chemistry. Anchor Book Doubleday & Company, Inc. Garden City, New York, 1965

Stengers, Isabelle and Bensaude-Vincent, Bernadette. A History of Chemistry. Harvard University Press. Cambridge, Massachusetts and London, England, 1996.

Berry, A.J. From Classical to Modern Chemistry. Some Historical Sketches. Cambridge At the University Press, New York, 1954.

Brock, William H. The Norton History of Chemistry. WW Norton & Company, New York, 1993.

Cobb, Cathy and Goldwhite, Harold. Creations of Fire Chemistry's Lively History from Alchemy to the Atomic Age. Plenum Publishing Corporation, New York, 1995.

Farber, Eduard. The Evolution of Chemistry A History of Its Ideas, Methods, and Materials. The Ronald Press Company, New York, 1952.

Greenberg, Arthur. A Picturing Chemistry Chemical from Alchemy to History Modern Molecular Science Tour. John Wiley & Sons, Inc. Public, Canada, 2000. Greenberg, Arthur. From Alchemy to Chemistry in

Picture and Story. John Wiley & Sons, Inc. New Jersey, 2007.

Hannaway, Owen. The Chemists and the Word The Didactic Origins of Chemistry. John Hopkins University Press, Maryland, 1975.

Marsh, J.E. The Origins and The Growth of Chemical Science. John Murray, Albemarle Street W, 1929.

Meyer, Ernst Von. A History of Chemistry from Earliest Times to the Present Day. Macmillan and Co., New York, 1891.

Morris, Richard. The Last Sorcerers The Path from Alchemy to the Periodic Table. Joseph Henry Press, Washington DC, 2003.

Muir, MM Pattison. A History of Chemical Theories and Laws. Arno Press, New York, 1975.

Partington, JR A History of Chemistry Volume One Part I: Theoretical Background. Macmillan & Co Ltd, New York, 1990.

Read, John. Humor and Humanism in Chemistry. G. Bell and Sons Ltd. London, 1947.

Partington, JR. A Short History of Chemistry 3rd Edition. Macmillan & Co Ltd. New York, 1965.

Salzberg, Hugh W. From Caveman to Chemist Circumstances and Achievements. American Chemical Society, Washington DC, 1991.

Schneer, Cecil J. Mind and Matter Man's Changing Concepts of the Material World. Grove Press, Inc. New York, 1969.

Stillman, John Maxson. The Story of Alchemy and Early Chemistry. Dover Publications, Inc. New York, 1960.

Soberer, KK The Goldmakers 10,000 Years of Alchemy. Greenwood Press Publishers, London, 1948.

Hopkins, Arthur John. AlchVemy Child of Greek Philosophy. AMS Press, Inc. New York, 1967.

Lindsay, Jack. The Origins of Alchemy in GraecoRoman Egypt. Frank Muller Ltd, 1970.

Morrison, Mark S. Modern Alchemy Occultism and The Emergence of Atomic Theory. Oxford University Press, 2007.

Pearsall, Ronald. The Alchemists. Weidenfeld and Nicolson Larlan.

Read, John. The Alchemist in Life, Literature, and Art. Thomas Nelson and Sons Ltd, New York, 1947.

Read, John. Prelude to Chemistry An Outline of Alchemy Its Literature and Relationships. G. Bell and Sons Ltd, London, 1939.

Redgrove, H. Stanley Alchemy: Ancient and Modern. University Books, Inc. New York, 1969.

Roberts, Gareth. The Mirrors of Alchemy Alchemical Ideas and Images in Manuscrpipts and Books from Antiquity to the Seventeenth Century. University of Toronto Press. North America, 1994.

Taylor, F. Sherwood. The Alchemists Founders of Modern Chemistry. William Heinemann Ltd. Great Britain, 1953.

Multhauf, Roberty P. The Origins of Chemistry. Franklin Watts, Inc. New York, 1967.

Moran, Bruce T. Distilling Knowledge Alchemy, Chemistry, and The Scientific Revolution. Harvard University Press, England, 2005.

https://en.wikipedia.org/wiki/Alchemy https://en.wikipedia.org/wiki/Chrysopoeia https://en.wikipedia.org/wiki/Elixir_of_life

https://en.wikipedia.org/wiki/Magnum_opus_(alchemy)

https://en.wikipedia.org/wiki/History_of_chemistry https://en.wikipedia.org/wiki/Ochre

https://en.wikipedia.org/wiki/Discovery_of_chemical_elements https://www.youtube.com/watch?v=BchgsTANO-k https://www.youtube.com/watch?v=0y8gpY1yhEU https://www.youtube.com/watch?v=aGjfa0UF608

https://www.youtube.com/watch?v=EWGsVzWV_i4&t=10s https://www.youtube.com/watch?v=bT3FpV2hOFc&t=522s https://www.youtube.com/watch?v=5qa4wolz_Ag https://www.youtube.com/watch?v=hhQC3glQkg4 https://www.youtube.com/watch?v=6_lgSaSMrQs

www.ingramcontent.com/pod-product-compliance
Lightning Source LLC
Chambersburg PA
CBHW031837200326
41597CB00012B/185

* 9 7 9 8 8 9 5 1 8 2 9 6 3 *